SUSTAINABLE AGRICULTURE

Second Edition

JOHN MASON

National Library of Australia Cataloguing-in-Publication:
Mason, John, 1951– .
Sustainable agriculture.
2nd ed.
ISBN 0 643 06876 7.
1. Sustainable agriculture. 2. Sustainable development.
I. Title.

338.16

Published by and available from:
Landlinks Press
PO Box 1139
Collingwood Vic. 3066
Australia

Telephone:	+61 3 9662 7666
Freecall:	1800 645 051 (Australia only)
Fax:	+61 3 9662 7555
Email:	publishing.sales@csiro.au
Website:	www.landlinks.com

Cover design and text design by James Kelly
Set in 10.5/13 Minion
Printed in Australia by BPA Print Group
Front cover photograph courtesy of John Mason

Contents

Acknowledgments

Research and Editorial Assistants: Iain Harrison, Peter Douglas, Paul Plant, Andrew Penney, Kathy Travis, Naomi Christian, Mark James, Alison Bundock, Rosemary Lawrence, Peter Douglas, Lisa Flower.

Thanks to the following organisations for information supplied:

The National Association for Sustainable Agriculture Australia (NASAA)

National Farmers Federation

Victorian Institute for Dryland Agriculture

Australian Wiltshire Horn Sheep Breeders Association

Australian Finnsheep Breeders Association

Llama Association of Australia

The Emu Producers Association of Victoria

The Australian Ostrich Association

Victorian Department Natural Resources and Environment

Introduction

First there was subsistence farming. Then there was a technological revolution: developments in machinery and chemicals allowed us to clear and cultivate land faster, feed plants and animals quicker (and grow them faster); and kill pests or diseases quickly. These new-found abilities seemed like a godsend to mankind; and throughout the 20th century we used them to their fullest, generally with little regard to any unforseen repercussions.

Gradually, time has revealed a variety of problems caused by this modern agricultural development, including chemical residues affecting plant and animal life on land and in the sea, soil degradation in the form of soil structural decline, erosion, salinity, soil acidification, loss of fertility, nutrient loading of waterways, dams and lakes and more.

As we move into the 21st century and concern about our environment grows, there is an obvious move towards more sustainable farming.

Sustainable farming is, in essence, concerned with anything that affects the sustainability of a farm. You cannot keep farming a property indefinitely if there is a degradation of resources (environmental resources, financial resources, equipment, machinery, materials, or any other resources). In the short to medium term, the problem of sustainability is overwhelmingly a financial one; but in the long term, environmental sustainability will possibly have a greater impact on the whole industry than anything else.

Why be sustainable?

If we can't sustain agricultural production, we will eventually see a decline in production; hence a decline in food and other supplies. There is no escaping the fact that people need agricultural products to survive: for food, clothing, etc. Science may be able to introduce substitutes (eg synthetic fibres) but even the raw materials to make these will generally be limited. As the world's population increases (or at best remains stable in some places) demand for agricultural produce increases accordingly. Poorly maintained farms produce less in terms of quantity and quality. Profitability decreases mean that surplus money is no longer available for repair and improvements. Farm land can become contaminated with chemical residues, weeds or vermin. The amount of vegetation produced (ie the biomass) may reduce, resulting in less production of carbon dioxide, and a greater susceptibility to environmental degradation.

We have created a world that relies heavily on technology to produce the food needed to sustain its human population. There is a worldwide dilemma. To abandon modern farming methods could result in worldwide famine but to continue current practices will almost certainly result in long-term degradation of farmland and, eventually, the inability to sustain even current human population levels, without even considering future increases in the human population.

Who should be concerned?

Everyone needs to be concerned about a decline in farm production potential. The farmer, his family, and workers are always affected first. An unsustainable farm is simply not worth persisting with and any farm which heads this way must eventually be abandoned or redeveloped to become sustainable. This book is about foreseeing and understanding such problems, and addressing them before it is too late.

1

Different things to different people

Sustainable farming means different things to different people, however they all share a common concern in preventing the degradation of some aspect of the farm. Some farmers are primarily concerned with degradation of natural resources (eg their land is becoming less productive). Other farmers may be more concerned about degradation of profitability, which could be due to increased labour or material costs, poor planning, or simply changing conditions in the economy. The causes and the solutions to such problems are different in each situation.

Sustainable agriculture is a philosophy: it is a system of farming. It empowers the farmer to work with natural processes to conserve resources such as soil and water, whilst minimising waste and environmental impact. At the same time, the 'agroecosystem' becomes resilient, self regulating and profitability is maintained.

What to do

There are many different ideas about how to be more sustainable. Different people promote different concepts with great vigour and enthusiasm and, in most cases, these concepts will contain something valuable. Many are quite similar in approach, often being variations of a similar theme. Each approach will have its application; but because it worked for one person does does not necessarily mean it will work for someone else. Some of these concepts are explained below.

Low input farming systems

This approach is based on the idea that a major problem is depletion of resources. If a farmer uses fewer resources (eg chemicals, fertiliser, fuel, money, manpower), farm costs will be reduced, there is less chance of damage being caused by waste residues or overworking the land, and the world is less likely to run out of the resources needed to sustain farming.

Regenerative farming systems

This seeks to create a system that will regenerate itself after each harvest.

Techniques such as composting, green manuring and recycling may be used to return nutrients to the soil after each crop. Permaculture is currently perhaps the ultimate regenerative system. A permaculture system is a carefully designed landscape which contains a wide range of different plants and animals. This landscape can be small (eg a home garden), or large (eg a farm), and it can be harvested to provide such things as wood (for fuel and building), eggs, fruit, herbs and vegetables, without seriously affecting the environmental balance. In essence, it requires little input once established, and continues to produce and remain sustainable.

Biodynamic systems

This approach concentrates on mobilising biological mechanisms. Organisms such as worms and bacteria in the soil break down organic matter and make nutrients available to pastures or crops.

Under the appropriate conditions, nature will help dispose of wastes (eg animal manures), and encourage predators to eliminate pests and weeds.

Organic systems

Traditionally this involves using natural inputs for fertilisers and pest control, and techniques such as composting and crop rotation. In Australia and many other countries, there are schemes which 'certify' produce as being organic. These schemes lay down very specific requirements, including products and farming techniques which are permitted, and others which are prohibited. In Australia, you can find out about such schemes through groups such as the Biological Farmers Association (BFA) or the National Association for Sustainable Agriculture (NASAA). See the Appendix for addresses.

Conservation farming

This is based on the idea of conserving resources that already exist on the farm. It may involve such things as, for example, identifying and retaining the standard and quality of waterways, creek beds, nature strips, slopes.

Hydroponics

This approach involves separating plant growth from the soil, and taking greater control of the growth of a crop. This increases your ability to manage both production and the disposal of waste.

Hydroponics is not a natural system of cropping, but it can be very environmentally friendly. A lot of produce can be grown in a small area; so despite the high establishment costs, the cost of land is much less, allowing farms to operate closer to markets. In the long term, a hydroponic farm uses fewer land resources, fewer pesticides, and is less susceptible to environmental degradation than many other forms of farming.

Matching enterprise with land capability

Some sites are so good that you can use them for almost any type of farming enterprise, for any period of time without serious degradation. Other places, however, have poor or unreliable climates or infertile soils and may only be suitable for certain types of enterprises or certain stocking or production rates. If you have a property already, only choose enterprises that are sustainable on your land.

(See the section on 'Assessing land capability' in this chapter.)

Genetic improvement

This principle involves breeding or selecting animal or plant varieties which have desirable genetic characteristics. If a particular disease becomes a problem, you select a variety that has reduced susceptibility. If the land is threatened with degradation in a particular way, you should change to varieties that do not pose that problem.

Polycultures

Many modern farms practise monoculture, growing only one type of animal or plant. With large populations of the same organism, though, there is greater susceptibility to all sorts of problems. Diseases and pests can build up to large populations. One type of resource (required by that variety) can be totally depleted, while other resources on the farm are under-used. If the market becomes depressed, income can be devastated. A polyculture involves growing a variety of different crops or animals, in order to overcome such problems.

Integrated management

This concept holds that good planning and monitoring the condition of the farm and marketplace will allow the farmer to address problems before they lead to irreversible degradation.

Chemical pesticides and artificial fertilisers may still be used, but their use will be better managed. Soil degradation will be treated as soon as it is detected. Water quality will be maintained. Ideally, diseases will be controlled before they spread. The mix of products being grown will be adjusted to reflect changes in the marketplace (eg battery hens and lot-fed animals may still be produced but the waste products which often damage the environment should be properly treated, and used as a resource rather than being dumped and causing pollution).

Know your land

Evaluating a site

Farmers need to know their property as well as possible, to ensure the best management decisions are made and the most suitable production systems and techniques are chosen.

Many site characteristics are seasonal so observations need to be made throughout the year, and over many years, to gain an ability to predict conditions. Changes to a site, such as removal or addition of vegetation in an area, can also alter future patterns.

The following are examples of useful measurements/indicators.

Weather patterns

Rainfall and temperature readings can help determine when to do different things (eg planting) and help plan future operations on a farm. Regional records do not show the subtle differences that can occur from one property to the next, or within different parts of the same property.

If possible keep your own records, but be sure to do so on a regular basis. Even a few weeks of missed records can give a distorted picture of local conditions.

Soil pH

This refers to how acidic or alkaline the soil is. Most pastures or crops have a preferred pH level in which to grow. Simple soil pH tests can allow you to change crops according to their suitability to different pH levels, or to carry out works to alter the soil pH to suit the crop you wish to grow. Failure to do so could result in expensive losses or greatly reduced yields. It is also important that tests are repeated at least every year or two, as pH levels can change over time, particularly if acidifying fertilisers are used, or the area has been regularly cropped with legumes.

Soil EC (electroconductivity)

An EC meter can be used to readily provide a quick reading of the electroconductivity of a soil sample. A higher EC reading indicates that electrons are flowing faster through the soil and indicates that there are probably more nutrients available to feed plants. Low readings indicate an infertile soil. Extremely high levels indicate toxic levels of chemicals in the soil (eg salinity).

Soil temperature

Use a portable temperature meter with a probe to measure at a depth of 10–15 cm. This enables farmers to determine when to sow (ie when germination temperatures are suitable for a crop or pasture species). Don't rely on one reading. Do several readings in different parts of the field/paddock to be seeded, as temperatures can vary from place to place. One high reading may give you a false outlook on the overall temperature conditions of the site.

Water conditions

The quality and quantity of water available will determine what crops or animals can be raised.

Some farming techniques make more efficient use of water than others (eg hydroponic produce may require less water than row cropping, but water quality must be excellent). Water quality may be gauged by simply performed measurements such as electroconductivity (EC) (see Chapter 4 for further information).

Monitoring soil moisture

Higher levels of nitrogen will bring an improved growth response in plants if soil is moist, but are wasted when soil is dry. It is useful to make two or more nitrogen applications to a broadacre crop (eg wheat), if and when moisture is appropriate. It is also important to pay attention to soil moisture at critical stages (eg sowing, tillering, flowering and pre-harvest). A neutron probe might be installed to make such measurements.

Electromagnetic characteristics

The electromagnetic characteristics of a site may indicate certain things about crop or livestock production capabilities, such as:

- Sources of underground water

- Natural radiation which can influence growth rates
- Sub-surface characteristics, such as certain mineral deposits

Factors affecting electromagnetic conductivity may include:

- Size of pores (porosity or spaces between soil particles)
- Amount of water between pores
- Soil temperature
- Salinity in soil and groundwater
- Mineral material in soil (eg clay, rock type)
- Amount of organic material

Electromagnetic characteristics of a soil can be measured by using a device such as an EM31 electromagnetic survey probe. It takes a degree of experience to use and interpret the results from such a probe, so be cautious about who advises you.

Herbicide or pesticide resistance

The effectiveness of certain chemicals can decline as weed or pest strains develop more resistance. It is valuable to ascertain if this is happening and change pesticide or weed control practices when resistance is seen, to ensure good control.

Land carrying capacity

A technique that is increasing in use classifies land into different types according to its characteristics. This can help determine potential for different uses. It aims to establish the best use for each land type, while hopefully balancing production (eg agriculture) versus other needs (eg conservation).

The characteristics of a site can affect:

- the type of enterprise it can be used for
- the type (quality and quantity) of inputs required to achieve different outcomes

Agricultural land in Australia is commonly classed into eight levels of capability or use, as shown in Table 1.

Table 1 Land Classes in Australia

Class	Description
I	Land suitable for all types of agriculture on a permanent basis
II	Land suitable for most types of agriculture on a permanent basis provided careful planning and simple modifications are applied (eg reduced tillage, fertiliser applications)
III	Arable land with moderate limitations for most types of agriculture provided careful planning and intensive management practices are applied
IV	Land with high levels of limitations, which usually requires high levels of management skill or it has low productivity
V	Very high limitations, low productivity and high management requirements
VI	Steep sloped or rocky land that is not traversable by standard equipment
VII	Extremely limited land which requires protection, productivity is not a significant factor
VIII	Land with no productive potential nor protection requirement

Source: *Land Care* by Bill Matheson (1996) Inkata Press

The use of these land types for agricultural production must be balanced against other required or potential uses for that land, including conservation, water catchment, etc.

Assessing land capability

The following steps can be used to assess land capability:

1. Draw plan(s) of farm property showing the characteristics of different areas (eg different paddocks) such as soil types, vegetation, drainage, etc.
2. Assess the capability of the land in different parts of the farm. You might need to regroup areas differently and rearrange current paddock divisions.
3. Determine management requirements in conjunction with proposed uses for different parts of the property.
4. Consider personal, financial, manpower and other resources to decide on land uses for different areas of the farm.

Consider the following criteria to categorise different parts of the property:

- Erosion potential
- Waterlogging/drainage (watertable)
- Soil pH
- Water repellence
- Soil fertility
- Soil structure
- Sub-soil structure
- Soil moisture-holding capacity
- Weather patterns
- Microclimate variations
- Existing vegetation

An indication of sustainability

Whether a farm is or is not judged to be sustainable will depend on the factors considered and the degree of importance attached to each factor. An Australian government committee (SCARM) in 1992 identified four key factors which they considered key indicators for sustainable agriculture. These are:

- Long-term real net farm income
- Land and water quality
- Managerial skills
- Off-site environmental effects and their attributes, as a basis for improved decision making at a national and regional level.

These indicators have been used a basis for ongoing research and planning in the development of sustainable agriculture in Australia.

2

Sustainable concepts

There are many suggested solutions to the sustainable farming problem. These range from 'landcare' and 'conservation farming' to 'permaculture', 'biodynamics' and 'financial restructuring'. Most of these solutions are very appropriate, in the right place and at the right time. All have their application and, in many cases, elements of several can be combined to create a solution appropriate to a particular site.

Natural farming

Natural farming works with nature rather than against it. It recognises the fact that nature has many complex processes which interact to control pests, diseases and weeds, and to regulate the growth of plants.

Chemicals such as pesticides and artificial fertilisers are being used more and more, even though they can reduce both the overall health of the environment and the quality of farm produce. Undesirable long-term effects such as soil degradation and imbalances in pest-predator populations also tend to occur. As public concern grows, these issues are seen as increasingly important. Farming the natural way aims to ensure quality in both the environment in which we live and in the produce we grow on our farms.

There are a variety of ways of growing plants that work with nature rather than against it. Some techniques have been used for centuries. Some of the most effective and widely used methods are outlined here.

Organic farming

Organic farming has been given a variety of names over the years – biological farming, sustainable agriculture, alternative agriculture, to name a few. Definitions of what is and isn't 'organic' are also extremely varied. Some of the most important features of organic

Figure 2.1 Ryton Organic Gardens. Vegetables growing at the Henry Doubleday Research Association grounds, United Kingdom.

production, as recognised by the International Federation of Organic Agriculture Movements (IFOAM), include:

- Promoting existing biological cycles, from micro-organisms in the soil to plants and animals living on the soil
- Maintaining environmental resources locally, using them carefully and efficiently and re-using materials as much as possible
- Not relying heavily on external resources on a continuous basis.
- Minimising any pollution, both on-site and leaving the site
- Maintaining the genetic diversity of the area

Typical practices used in organic systems are composting, intercropping, crop rotation and mechanical or heat-based weed control. Pests and diseases are tackled with naturally produced sprays and biological controls (eg predatory mites). Organic farmers generally avoid the use of inorganic (soluble) fertilisers, synthetic chemical herbicides, growth hormones and pesticides.

One of the foundations of organic farming, linking many other principles together, is composting. By skilfully combining different materials, balancing carbon and nitrogen levels, coarse and fine ingredients, bacteria and worms act to break down the waste products. Composting produces a valuable fertiliser that can be returned to the soil. Natural biological cycles are promoted, 'wastes' are re-used and the need for external supplies of fertiliser are reduced or cut altogether (see Chapter 3 for more information on composting).

Whole farm planning

This concept encourages a 'holistic' and long-term approach to farm planning. It requires giving due consideration to all of the farm assets (physical and non physical) over a long period of time (perhaps several generations); with respect to all of the aims which the farmer may aspire to (eg profit, lifestyle, family wellbeing, sustainability of production).

In any whole farm planning strategy the farmer must first assess the site in terms of potential use/suitability. The farm is then subdivided, usually by fences, to emphasise useful or problem areas (eg erosion, salinity). Water and access routes are highlighted. Cropping or livestock rates are planned to be increased if feasible. Shelter is planned and planted out, or built. Pest animals and plants are located, identified and controlled by chemical or natural alternatives.

Conservation is a very important aspect of whole farm planning; native birds and animals are mostly beneficial on the farm as they control a range of pest animals and insects.

Costs inevitably will be a deciding factor. The farmer needs to determine what costs may be involved and what the benefits of whole farm planning are to the future of the farm. In the majority of cases, long-term gains far outweigh the time and resources used in establishing such a plan. Information on whole farm planning is readily available from agencies such as state government departments of agriculture, primary industries, conservation or land management.

Systems thinking in sustainable agriculture

The role of the farmer in a systems or holistic farm approach to agriculture is to organise and monitor a whole system of interactions so that they keep one another in shape. The farmer is interested not only in producing the maximum amount of the species that he draws his income from, but also in minimising inputs such as chemicals, fertilisers and cultivations that cost money. Such systems are more sustainable in the long term. Whilst the overall production of many sustainable farms may be lower, the cost of inputs is also lower, meaning that overall profit is still comparable to conventional systems.

Permaculture

In its strictest sense, permaculture is a system of production based on perennial, or self-perpetuating, plant and animal species which are useful to people. In a broader context, permaculture is a philosophy which encompasses the establishment of environments which are highly productive and stable, and which provide food, shelter, energy, etc., as well as supportive social and economic infrastructures. In comparison to modern farming techniques practised in Western civilisations, the key elements of permaculture are low energy and high diversity inputs. The design of the landscape, whether on a suburban block or a large farm, is based on these elements.

A permaculture system can be developed on virtually any type of site, though the plants selected and used will be restricted by the site's suitability to the needs of the varieties used. Establishing a permaculture system requires a reasonable amount of pre-planning and designing. Factors such as climate, landform, soils, existing vegetation and water availability need to be considered. Observing patterns in the natural environment can give clues to matters which may become a problem later, or which may be beneficial.

A well designed permaculture farm will fulfil the following criteria:

- Upon maturity it forms a balanced, self-sustaining ecosystem where the relationships between the different plants and animals do not compete strongly to the detriment of each other. Althought the farm does not change a great deal from year to year, nonetheless it still continues to change.
- It replenishes itself. The plants and animals on the farm feed each other, with perhaps only minimal feed (eg natural fertilisers) needing to be introduced from the outside.
- Minimal, if any, work is required to maintain the farm once it is established. Weeds, diseases and pests are minimal due to companion planting and other natural effects which parts of the ecosystem have on each other.
- It is productive. Food or other useful produce can be harvested from the farm on an ongoing basis.
- It is intensive land use. A lot is achieved from a small area. A common design format used is the Mandala Garden, based on a series of circles within each other, with very few pathways and easy, efficient watering.
- A diverse variety of plant types is used. This spreads cropping over the whole year, so that there is no time when a 'lot' is being taken out of the system. This also means that the nutrients extracted (which are different for each different type of plant or animal) are 'evened out' (ie one plant takes more iron, while the plant next to it takes less iron, so iron doesn't become depleted as it would if all the plants had a high demand for iron). The diversity of species acts as a buffer, one to another.
- It can adapt to different slopes, soil types and other microclimates.
- It develops through an evolutionary process changing rapidly at first but then gradually over a long period – perhaps never becoming totally stable. The biggest challenge for the designer is to foresee these ongoing, long-term changes.

Structure of a permaculture system

- Large trees dominate the system. The trees used will affect everything else – they create shade, reduce temperature fluctuations below (create insulation), reduce light intensities below; reduce water loss from the ground surface, act as wind barriers, etc.
- In any system, there should also be areas without large trees.
- The 'edge' between a treed and non-treed area will have a different environment to the areas with and without trees. These 'edges' provide conditions for growing things which won't grow fully in the open or in the treed area. The north edge of a treed area (in the southern hemisphere) is sunny but sheltered while the south edge is cold but still sheltered more than in the open. 'Edges' are an example of microclimates, small areas within a larger site that have special conditions which favour certain species which will grow well elsewhere (see also the section on corridor planting in Chapter 9 for more information on 'edge' effects).
- Pioneer plants are used initially in a permaculture system to provide vegetation and aid the development of other plants which take much longer to establish. For example, many legumes grow fast and fix nitrogen (raise nitrogen levels in the soil) and thus increase nutrients available to nut trees growing beside them. Over time the nuts will become firmly established and the legumes will die out. Pioneer plants

are frequently (but not always) short lived.

The concept of permaculture was developed by Bill Mollison of Australia.

Minimal cultivation

Cultivation of soil is often used extensively in organic growing, particularly to control weed growth. Where chemical weedicides are not used, ploughing or hoeing can be extremely effective methods of controlling weeds. These techniques also help to open up soils which have become compacted, allowing water and air to penetrate more readily into the soil. Cultivation has been shown (by ADAS research, UK) to help reduce plant disease by destroying plants which might harbour those diseases.

There are problems with cultivation, however, as outlined below:

- It can destroy the soil profile, the natural gradation from one type of soil at the surface (usually high organic and very fertile) through layers of other soil types as you go deeper in the soil. When the soil profile is interfered with, hard pans can be created. A pan is a layer beneath the surface of the soil where water and root penetration becomes difficult. Water can build up over a hard pan creating an area of waterlogged soil.
- Drainage patterns can be changed
- Plant roots can be damaged
- Heavy machinery can cause compaction
- Shallow cultivation can encourage weed seed germination. Cultivation can also bury seed and protect it from foraging birds and rodents; it may also help keep it moist and warm enough to germinate
- Loosened soil can be more subject to erosion (eg from wind, rain, irrigation)

Figure 2.2 No dig garden with compost bin at rear.

No dig techniques

Techniques where the soil is not dug or cultivated have some obvious advantages. Some of the techniques used in this approach are pest, disease and weed control with fire, mulching for weed control and water retention, and raised organic beds.

Vegetable-sod inter planting

This involves growing mulched rows of vegetables 20–40 cm wide over an existing mowed turf. A narrow line may be cultivated, sometimes down the centre of each row, to sow seed into, if growing by seed, to hasten germination. Mulch mats, black plastic, paper or organic mulches can be used to contribute to weed control in the rows. Crop rotation is usually practised between the strips. This contributes towards better weed control. Clover is often encouraged in the strips of turf between rows to help improve nitrogen supplies in the soil.

No dig raised beds – one method

Build four walls for each bed from timber. Use a wood which will resist rotting such as red gum, jarrah, recycled railway sleepers or even treated pine. The dimensions of the box can be varied but commonly might be 20–30 cm or more high and at least 1 m wide and 1–3 m or more long. The box can be built straight on top of existing ground, whether pasture, bare earth or even a gravel path. There should be a little slope on the ground it is built over to ensure good drainage. It may also be necessary to drill a few holes near the base of the timber walls to ensure water is not trapped behind them. Weed growth under and around the box should be cleaned up before it is built. This may be done by burning, mowing, hand weeding, mulching, or a combination of techniques.

The box can be filled with good quality organic soil, compost, or some other soil substitute such as alternate layers of straw and compost from the compost heap or alternate layers of graded and composted pine bark, manure and soil. The growing medium must be friable, able to hold moisture, and free of disease and weeds (avoid materials, such as grass hay, or fresh manures that may hold large quantities of weed seeds).

A commonly used watering technique in these beds is to set a 2 L plastic bottle (eg soft drink or milk) into the centre of the bed below soil level. Cut the top out, and make holes in the side. This can be filled with water, which will then seep through the holes into the surrounding bed. Mulching the surface may be desirable to assist with controlling water loss and reducing weeds (Reference: *Organic no dig, no weed gardening* by Pincelot, published by Thorsons).

Biodynamics

Biodynamic farming and gardening is a natural practice which developed from a series of lectures given by Rudolf Steiner in 1924. It has many things in common with other forms of natural growing, but it also has a number of unique characteristics.

Biodynamics views the farm or garden as a 'total' organism and attempts to develop a sustainable system where all of the components of the living system have a respected and proper place.

There is a limited amount of scientific evidence available which relates to biodynamics. Some of what is available suggests biodynamic methods do in fact work. It will, however, take a great deal more research for mainstream farmers to become convinced of the effectiveness of these techniques; or in fact for the relative effectiveness of different biodynamic techniques to be properly identified.

Principles of biodynamics

- Biodynamics involves a different way of looking at growing plants and animals.
- Plant and animal production interrelate; manure from animals feeds plants and plant growth feeds the animals.
- Biodynamics considers the underlying cause of problems and attempts to deal with those causes rather than treating problems in a superficial way. Instead of responding to poor growth in leaves by adding nutrients, biodynamics looks at what is causing the poor growth – perhaps soil degradation or wrong plant varieties – and then deals with the bigger question.
- Produce is a better quality when it is 'in touch' with all aspects of a natural ecosystem. Produce which is produced artificially (eg battery hens or hydroponic lettuces) will lack this contact with 'all parts of nature' and, as such, the harvest may lack flavour, nutrients, etc. and not be healthy food.
- Economic viability and marketing considerations affect what is grown
- Available human skills, manpower and other resources affect what is chosen to be grown
- Conservation and environmental awareness are very important
- Soil quality is maintained by paying attention to soil life and fertility
- Lime, rock dusts and other slow-acting soil conditioners may be used occasionally
- Maintaining a botanical diversity leads to reduced problems
- Rotating crops is important
- Farm manures should be carefully handled and stored
- Biodynamics believes there is an interaction between crop nutrients, water, energy (light, temperature) and special biodynamic preparations (ie sprays) which result in biodynamically produced food having unique characteristics.
- Plant selection is given particular importance. Generally, biodynamic growers emphasise the use of seed which has been chosen because it is well adapted to the site and method of growing being used.
- Moon planting is often considered important. Many biodynamic growers believe better results can be achieved with both animals and plants if consideration is given to lunar cycles. They believe planting, for example, when the moon is in a particular phase; can result in a better crop.

Developing a biodynamic farm or garden

The first step is always to look at the property as a single organism and to appreciate that whatever changes are made to the property can have implications to many (probably all) of the component parts of that property. There is an obvious (though sometimes subtle) relationship between every plant or animal and its surroundings – both nearby and distant.

Biodynamic preparations/sprays

These are a unique and important aspect of biodynamics. There are all sorts of biodynamic preparations and wide experience (throughout many countries) suggests the use of these preparations is beneficial, resulting in both morphological and physiological changes in plants (eg better ripening rates, better dry matter, carbohydrate and protein rates).

Some of these special preparations are outlined below:

- In *Organic Farming* by Lambkin (Farming Press, UK) two different sprays (500 and 501) are mentioned as being commonly used. These are made from a precise formulation of quartz and cow manure and are sprayed on crops at very diluted rates. Biodynamic growers in the UK and elsewhere also use preparations made from plants to stimulate compost and manure heaps.
- Cow manure is placed in a cow horn and buried over winter, with the intention of maintaining a colony of beneficial organisms in the horn over the cold months which can then recolonise the soil quickly in the spring.
- Insect control sprays are commonly made as follows. Catch some of the grubs or insects which are becoming a pest. Mash them to a pulp (perhaps in a food processor), then add water and place in a sealed jar for a few days in a refrigerator. Once fermentation begins, remove and dilute with water (100:1). Spray over affected plants. This is said to repel the insects, though no scientific evidence is known to support the treatment.
- Biodynamic growers use a variety of different preparations to add to compost heaps or spray on paddocks or garden plots to encourage faster decomposition. Preparations have included: yarrow flowers, valerian flowers, oak bark, calendula flowers, comfrey leaves and preparations from Casuarina and Allocasuarina species.

Crop rotation

Crop rotation consists of growing different crops in succession in the same field, as opposed to continually growing the same crop. Growing the same crop year after year guarantees pests of a food supply – and so pest populations increase. It can also lead to depletion of certain soil nutrients. Growing different crops interrupts pest life cycles and keeps their populations in check. Crop rotation principles can be applied to both broadacre and row crops alike. The principles may even be applied to pastures.

In the United States, for example, European corn borers are a significant pest because most corn is grown in continuous cultivation or in two-year rotations with soybeans. If the corn was rotated on a four or five-year cycle, it is unlikely that corn borers would be major pests. This kind of system would control other corn pests as well as corn borers.

In crop rotation cycles, farmers can also sow crops like legumes that actually enrich the soil with nutrients, thereby reducing the need for chemical fertilisers. For example, many corn farmers alternate growing corn with soybeans, because soybeans fix nitrogen into the soil. Thus, subsequent corn crops require less nitrogen fertiliser to be added.

Crop rotation in vegetables

Look at the list of groups of vegetables below. Don't grow a vegetable in a particular area if another vegetable from the same group was grown in that spot recently. Keep varying the type of vegetable grown in a particular spot. Crop rotation can also include a fallow period, when a non-harvested crop is grown.

- Brassicas (formerly Cruciferae): broccoli, brussels sprouts, cabbage, cauliflower, sea kale, kohl rabi, turnip, swede, radish, horseradish etc.

- Cucurbitaceae: cucumber, marrow, pumpkin, squash, cantaloupe (ie rock melon), zucchini
- Liliaceae: onion, leeks, garlic, asparagus, chives
- Fabaceae (legumes): peas, beans, clover
- Poaceae: corn, other grains
- Apiaceae (formerly Umbelliferae): celery, carrot, parsnip, fennel
- Asteraceae (formerly Compositae): chicory, lettuce, endive, globe artichoke, sunflower
- Chenopodiaceae: silver beet, red beet (ie beetroot) and spinach
- Solanaceae: tomato, capsicum, potato, eggplant

Seed saving

When plants are allowed to naturally pollinate each other, produce flowers, fruit and then seed, the local conditions will determine whether the offspring of those plants are suitable for the area. Plant varieties that have been bred in another state or country may not be suited to a different locality without large inputs of fertilisers or pesticides. Growing your own herbs and vegetables, for example, can provide the ideal seed source for your conditions.

Only collect seed from healthy plants, preferably with good yields and pleasant tasting produce. Wait until the seeds are ripe before harvesting, although be careful not to let all the seed fall out or blow away. Seeds should usually be stored in paper bags or envelopes, and kept in cool, dry and dark conditions. It is helpful to label your seeds with species, place grown, time harvested, etc.

Hydroponics

Hydroponics is a process used to grow plants without soil. As such, the grower is taking 'control' of the plant's root environment, and losing the benefit of 'mother nature's' finely tuned mechanisms which normally control that part of the plant's environment.

Hydroponics is not an easier way to grow things, but it is a more controlled way of growing plants. Growing in hydroponics can offer the following advantages:

- It can reduce the physical work involved in growing
- It can reduce the amount of water used in growing
- It can allow more efficient use of inputs such as fertiliser and pesticide, hence significantly less chemical is used
- It can allow a greater control of waste product, thus eliminating, or at least reducing, soil degradation or other forms of environmental damage
- It can save on space ... more can be grown in the same area

When you remove the soil from a plant and take control of its roots, it is essential that you have a good understanding of how it grows. Anybody can grow plants in soil with reasonable success, because nature is at work buffering your mistakes BUT, to grow plants in hydroponics you must understand how the plant grows so that you can control the temperature, water, oxygen, nutrients, etc. in the root zone.

Environmentally friendly farming

There are a number of other ways in which you can go about your farming that are environmentally friendly.

More efficient engines

Even if it costs more initially, it will pollute less, and cost less to operate, providing long-term savings.

- Keep engines running well and clean. Regularly remove wet grass or material that wraps around moving parts. Regularly carry out maintenance requirements, such as cleaning air filters, particularly in dusty conditions. Replace worn or damaged parts.
- Make sure you use the right sized engine for the job. Too small and it will be under strain, causing the engine to run inefficiently and to quickly wear out, requiring its repair or replacement. Too large and you are wasting fuel and probably making a lot more noise than is necessary. It is a good idea to get advice from a reputable distributor of power products.
- Performance products such as corrosion inhibitors and friction modifiers will often improve engine efficiency.

Alternatives to petrol engines

The exhausts from engines used to propel machinery such as tractors, mowers, chainsaws, etc contribute to air and noise pollution. Where possible, try to use methods that don't require petrol engines. For example:

- Instead of using petrol power, use cleaner energies such as electricity.
- For small jobs, use shears or a scythe instead of a brush cutter, use a hand saw or axe instead of a chainsaw. These tools work well if the blade is sharp!
- Grazing animals such as sheep or goats can be used to keep grass and weeds under control.

Burning off

Try to avoid burning off at anytime. As a general rule, if you can burn it, you can probably compost or recycle it. This means you don't waste useful material and you don't pollute the atmosphere or upset your neighbour when the wind blows smoke or ashes into their property. Many local councils now ban or strictly control any burning off, whether in the open or in incinerators.

Utilising energy produced on your property

There are many energy sources that can be utilised, including:

- Windmills for pumping water or generating electricity
- Solar panels used to generate electricity
- Solar energy collectors used to provide heating for animals, plants or the farm house
- Biogas generated from composting or decomposition of waste products such as manure, sewage, wood chips, mulch, or spoiled hay.

Checklist of sustainability elements

When you purchase a property, check for evidence that the property will be conducive to sustainable farming practices. Check the following elements initially, and monitor them after purchase:

Soil chemical characteristics

- Salt build up/residues of unwanted chemicals
- Soil pH

Soil physical characteristics

- Erosion
- Organic content
- Compaction
- Drainage problems

Soil life

- Microbes, earthworms etc
- Bad pathogens and microbes

Water supply

- Becoming depleted (groundwater being overused, watercourses being diverted, dammed, pumped upstream, etc) irrigation schemes losing funding, becoming under-maintained
- Access to water storage becoming more costly
- Changing weather patterns (drought etc)
- Contamination

Pest populations

- Weeds
- Animal pests (eg rabbits)
- Insects
- Fungal diseases

Sustainable agriculture around the globe

The widespread development of low input agricultural systems depends not only on the desires of farmers and consumers, but also upon national and international policy changes. Many existing policies favour high input – high output agricultural systems. However, governments around the world have begun to recognise the need for sustainable agricultural practices.

In 1972 the US government established the Integrated Pest Management (IPM) Program which aims to decrease the use of chemical pesticides by teaching farmers how to use a variety of biological controls, genetic resistance, and appropriate use of tillage, pruning, plant density and residue management. In 1977, the US government developed 'best management practices' (BMPs) including the use of cover crops, green manure crops, and strip cropping to minimise erosion; soil testing and targeting and timing of chemical applications to prevent the loss of nutrients and pesticides. District officers use these BMPs to help farmers develop conservation plans for their farms. The Agricultural Conservation Program provides funding for farmers to commence conservation practices such as crop rotation, biological pest control, soil testing and ridge tilling. Currently, the US Government has a Sustainable Agriculture Research and Education Program (SARE) and an Integrated Farming Systems Program. These programs point to a greater commitment to sustainable agricultural principles in the future.

The Australian government has acknowledged the need for 'sustainable development of agricultural industries' to 'contribute to' long-term productivity, and to Australia's 'economic wellbeing'. In addition, it acknowledges the need to protect the biological and physical resources which agriculture depends upon.

A strategic approach has been developed requiring cooperative action from different agencies, all levels of government, community and agricultural industries, across Australia. This approach has put forward five objectives as follows:

1 Create a framework of integrated government policies and programs which promote community based self reliant approaches to agricultural resource management.
2 Promote integrated planning of agricultural resource management, particularly in areas affected by land degradation; and extend measures (particularly community based self help approaches) which encourage information transfer and landholder adoption of sustainable management.
3 Reduce and manage effectively the impacts of pest plant and animal species on Australia's agricultural areas.
4 Improve kangaroo management at the national level, including removal of impediments to a sustainable commercial kangaroo industry.
5 Improve effective and safe management of agricultural and veterinary chemicals while improving levels of, and access to, information on these chemicals.

In the United Kingdom, the government has established the Sustainable Development Commission. The commission's role is to advocate sustainable development across all sectors in the UK, review progress towards it, and build consensus on the actions needed if further progress is to be achieved. The Sustainable Development Commission has published 'A Vision for Sustainable Agriculture'. It states that agriculture must:

- Produce safe, healthy food and non-food products in response to market demands, not only now, but in the future
- Allow producers to earn livelihoods from sustainable land management, taking account of payments for public benefits provided
- Operate within biophysical and environmental constraints
- Provide benefits such as environmental improvements to a public that wants them
- Maintain the highest standards of animal health and welfare compatible with society's right of access to food at a fair price
- Support the strength of rural economies and the diversity of rural culture

- Sustain the resources available for growing food and supplying other public benefits over time, except where alternative land uses are essential in order to meet other needs of society.

In developing countries, as opposed to developed countries, a large proportion of the population is engaged in farming activities. The conventional agricultural practices in developed countries are designed to minimise a scarce resource: labour. This is achieved by using pesticides, chemical fertilisers and heavy machinery, where manual labour would be used in developing countries. When these kinds of farming systems are moved from wealthy to poor countries, the results can be devastating. For example, developing countries often have limited space available for cultivation, and the soil in many countries is not very fertile to begin with. When cultivation techniques further degrade the soil, it becomes less useful for cultivation. Farmers notice this loss of production and move to a different spot, leaving the nutrient-poor soil to turn into wasteland. In some cases fertiliser is overused, causing soil degradation. The excess fertiliser can contaminate groundwater, as does pesticide residue.

A lack of education and regulation mean that pesticides are sometimes overused in developing countries. Poor irrigation practices are also indicative of a lack of research and education. It is becoming increasingly evident that conventional agriculture is not a long-term option in developed or developing countries. While developed countries are beginning to recognise and make policy decisions which recognise the importance of sustainable agriculture, many developing nations are not afforded this luxury. Whether sustainable agricultural processes are a viable proposition in developing countries is the subject of much current research.

Target 10: A model for sustainable agricultural development

Target 10 is a dairy industry project that has operated in Victoria, Australia, since 1992. It is the initiative of the Department of Primary Industries (Victoria), United Dairyfarmers of Victoria and the Dairy Research and Development Corporation and is supported by related industry bodies, and research and tertiary institutions. Initially, the program was mainly concerned with improved pasture consumption as the most likely opportunity on the dairy farm for increased profitability. The quality and quantity of milk and butterfat being produced are directly related to the quality and quantity of feed the cows are consuming.

The program now focuses on grazing management and other areas of high priority to the industry such as soils and fertilisers, cow nutrition, dairy farm performance analysis, natural resource management, empoyment and business planning.

The program has been enthusiastically accepted by Victorian dairy farmers and a recent survey shows that around 80% of farmers believe they have or will benefit from the program. Over 40% of farmers have already participated in one or more of the core programs.

Target 10 is immediately concerned with the attitudes of the dairy farming community. Field days, work groups and numerous other methods are employed by the Target 10 co-ordinators to make use of relevant information. Meetings are coordinated to be held prior to major decision making periods of the farming year. In this way discussion tends to be topical and any learning that is absorbed by the individual can be applied while the ideas are still fresh. In addition, follow up consultancy and advice are readily available to those farmers who are less comfortable with changing methods that may have been in use on the family farm for decades.

The program is well coordinated and provides significant profitability for those involved. While it is primarily dairy orientated, it would serve as an excellent model for similar programs in other agricultural and horticultural enterprises.

The program has a website hosted by the Department of Primary Industries, at www.target10.com.au. The website provides up-to-date news on the dairy industry, as well as access to a range of resources including on-line manuals and interactive decision support systems.

3

Soils

Introduction

The thin layer of soil at the earth's surface is essential for maintaining life. This layer of soil is the basis of most agriculture around the globe. If soil is lost or degraded, the potential of an area to support both plant and animal life is greatly reduced. Whilst the actual definition of sustainable farming varies somewhat, conservation and rejuvenation of agricultural soils are essential elements of any sustainable agricultural system. It is important to understand that soil is not the property of the land owner, the lease holder or the tenant of the site. It is the property of everyone, now and in the future. It may take many thousands of years for a soil to form, but only a few years for it to be degraded or lost due to poor management practices. For this reason it is critical that the techniques we use to manage our soils will maintain them in a manner that ensures that they are at least as productive for future generations as they are now, and hopefully are even improved.

Growing media

The growing medium is the material (or space) in which plant roots grow. This has traditionally been soil, but with the application of modern technology, we are provided with other options for growing media. There are three main options:

1. Sustained Organic Soils. In nature, the best soils contain at least 5–10% organic matter. Organic matter influences soil and water systems which in turn modify organic matter turnover and nutrient cycling. Maintaining the levels of organic matter in soils is a vital element of sustainable agricultural production.
2. Technologically Supported Soils. By adding fertilisers and soil ameliorants (eg lime) and by irrigating heavily, technology enables us to grow almost anything in almost any soil. Unfortunately, this is an inefficient use of resources and can result in serious degradation of soils, as well creating further problems, such as excess nutrients entering our waterways and causing problems such as algal blooms.

3. Hydroponics. Despite being an 'artificial' way of growing plants, hydroponics can be sustainable if it is well managed and relatively environmentally friendly. It allows you to take full control over the root zone:
   - excess nutrients are not 'lost' and washed into surrounding areas
   - chemical residues can be collected, treated and disposed of properly
   - water loss can be minimised, and water use maximised

Of the three options above, the least sustainable is the most commonly practised (ie technologically supported soils). This is probably because it is the easiest approach in the short term to overcome existing problems. However, what usually occurs is that this method develops its own problems.

Soils

Plant growth is directly affected by the type of soil the plants are grown in. The majority of plants depend on soil to provide nutrition, physical support (ie a place for roots to anchor), water and air. The exceptions to this are those plants known as epiphytes. These grow in such places as tree trunks, on rocks, or on fallen logs. The quality and quantity of plant growth will also affect how well other organisms (eg grazing animals, humans) will survive.

Soil is composed of:

1 Particles, which consist of:
   - Mineral particles of various sizes including clays, silts and sands
   - Organic material in varying states of decomposition
   - Living organisms – mostly microscopic, but also including insects, earthworms, nematodes, etc.

2 Water, which contains varying amounts and types of nutrients (and other chemicals) in solution

3 Air

These things affect the soil's ability to grow plants. It is possible to grow some plants in soils without living organisms, organic matter or mineral particles but plant roots must have air, water and nutrients. Generally however, you will require some amount of each of the above components to get the best growth from your plant.

The best types of soil for most agricultural production can be described as having the following attributes:

- well drained
- deep root zone
- easily penetrated by air, water and roots
- good water-holding capacity
- maintains a balanced nutrient supply
- erosion resistant.

Where these attributes do not exist or are in need of aid, then intervention from, for example, the farmer or land owner, is required to carry out management practices that will improve these characteristics, and ensure their sustainability.

Problems with soils

Soils are constantly affected by all that goes on around them, both by natural processes, and particularly in recent history, by human activity. Some of the more common human-induced changes include:

- Reduction in vegetation cover, opening up soils to increased rates of erosion
- Loss of soil structure due to poor cultivation techniques and the passage of heavy equipment, or regular traffic by hoofed animals (eg cows, horses)
- Reduction in soil fertility by not replacing nutrient losses from agricultural production
- An increase in saline-affected soils due to vegetation clearance or poor irrigation practices
- Waterlogging due to poor irrigation practices
- Soil acidification through the extensive use of acidifying fertilisers, or the extensive cropping of plants that remove large amounts of calcium from the soil (eg lucerne)
- Pollution of some soils through the use of persistent or toxic agricultural chemicals

When plants (trees and shrubs) are cleared from a site, soil is exposed to sunlight and the eroding effects of wind and water. Soil aeration is increased and the rate of weathering increases.

Apart from erosion, the proportion of organic matter in the soil gradually decreases through the action of microbes in the soil which use it as a source of energy – unless the new land use provides some replacement.

Major types of soil problems

Major soil problems include:

1 Loss of soil fertility
2 Erosion
3 Salinity
4 Soil sodicity
5 Soil structural decline
6 Soil acidification
7 Build up of chemical residues

Loss of soil fertility

Sustainable soil fertility implies that soil nutrients will be available in the same quantity in the long term. For this to happen, nutrients that are removed from the soil (in the form of plant crops, livestock, leached nutrients, cleared vegetation etc) need to be replaced. In a sustainable system, farm managers strive to work with the natural nutrient cycle to ensure that nutrients do not build up or get lost. The main problem in terms of agricultural production is a loss of soil nutrients. There are a number of ways to reduce the loss of nutrients from the soil, including applying manure produced from the soil back to it and minimising losses due to erosion, denitrification and leaching.

Minimising denitrification

Soil denitrification (ie loss of nitrogen to the atmosphere) is affected by the nature and quantity of organic matter present, degree of soil aeration, soil moisture content, pH and temperature. Management practices can reduce nitrogen loss caused by these factors. For example, while farmers may have little control over rainfall, they can manipulate soil moisture content through irrigation and drainage practices. Saturated soils can produce the anaerobic environment required for denitrification, so irrigation practices can be timed to avoid waterlogging.

Adding large amounts of organic matter to soils can result in high rates of microbial expansion, which in turn leads to an anaerobic environment which denitrifies quickly. This presents a dilemma, as organic matter is a highly beneficial ingredient in sustainable soil systems. In the long term, it increases aeration which improves soil structure, so making denitrification less likely.

Organic nutrients should be applied as close as possible to the time when they will be required by the crop. If it is essential to apply the nitrogen ahead of time, planting a cover crop that will accumulate the nitrogen may be necessary. The nitrogen can be stored and released for future use by decomposition.

Nitrogen should be applied in an even concentration over the entire area. Localised buildups will accelerate denitrification.

Minimising loss of nutrients due to leaching

To minimise loss due to leaching, it is advisable to only supply nutrients at a rate equal to the rate of uptake by the crop. In highly permeable soils, this can take the form of several applications of fertiliser as opposed to one large hit at the beginning of the growing cycle.

Green manure crops that are ploughed back into the soil will reduce nitrogen as they decompose. However, green manures that produce more nitrogen than the crop needs will result in nitrogen leaching out of the root zone. The rate of release of nitrogen from cover crops can be slowed by leaving the plant on the surface of the soil rather than digging it in.

Erosion

Soil erosion, which is the movement of soil particles from one place to another by wind or water, is considered to be a major environmental problem. Erosion has been going on through most of earth's history and has produced river valleys and shaped hills and mountains. Such erosion is generally slow but the action of humans has caused a rapid increase in the rate at which soil is eroded (ie a rate faster than natural weathering of bedrock can produce new soil). This has resulted in a loss of productive soil from crop and grazing land, as well as layers of infertile soils being deposited on formerly fertile crop lands; the formation of gullies; silting of lakes and streams; and land slips.

Common human causes of erosion

- Poor agricultural practices, such as ploughing soil that is too poor to support cultivated plants or ploughing soil in areas where rainfall is insufficient to support continuous plant growth
- Exposing soil on slopes
- Removal of forest vegetation
- Overgrazing – removing protective layers of vegetation and surface litter

- Altering the characteristics of streams, causing bank erosion
- Causing increased peak water discharges (increased erosion power) due to changes in hydrological regimes, by such means as altering the efficiency of channels (channel straightening); reducing evapotranspiration losses as a consequence of vegetation removal; and by the creation of impervious surfaces such as roads and footpaths, preventing infiltration into the soil and causing increased runoff into streams

Types of erosion

The two basic types of erosion are:

1. Water erosion
2. Wind erosion

Water erosion

With water erosion, soil particles are detached either by splash erosion (caused by raindrops) or by the effect of running water. Several types of water erosion are common in our landscapes. These are:

1. Sheet erosion – where a fairly uniform layer of soil is removed over an entire surface area. This is caused by splash from raindrops, with the loosened soil generally transported in rills and gullies.
2. Rill erosion – occurs where water runs in very small channels over the soil surface, with the abrading effect of transported soil particles causing deeper incision of the channels into the surface. Losses consist mainly of surface soil.
3. Gully erosion – occurs when rills flow together to make larger streams. They tend to become deeper with successive flows of water and can become major obstacles to cultivation, and a major threat to the safety of livestock. Gullies only stabilise when their bottoms become level with their outlets.
4. Bank erosion – is caused by water cutting into the banks of streams and rivers. It can be very serious at times of large floods and cause major destruction to property.

Wind erosion

Wind erosion can quickly remove soil particles, including nutrients and organic material, from the soil surface (generally the most fertile part of the soil). Dislodged soil can be lost totally from a property, or may be deposited in places where it is a problem, such as against fences or walls, on roadways, in clothing, washing, or inside the house. Windblown particles can also damage plants (ie sandblasting effect).

The force of wind becomes strong enough to cause erosion when it reaches what is known as the 'critical level' and this is the point at which it can impart enough kinetic energy to cause soil particles to move. Particles first start rolling along the surface. Once they have rolled a short distance they often begin to bounce into the air, where wind movement is faster. The effect of gravity causes these particles to fall back down to the surface where they either bounce again or collide with other particles. This process is known as 'saltation'.

Two other ways of windborne particle movement occur. The first is 'free flight', which occurs where very small particles are transported in air. The air acts as a fluid and carries them long distances. The other is called 'surface creep', where soil particles are too large to bounce and are rolled downwind.

As wind moves across an even layer of soil, the wind speed is relatively minor close to the soil (and up to 1 inch or 2.5 cm above the soil); above that there is a smooth flow of air; and above that air flow is turbulent. If the soil is uneven, anything projecting into the turbulent layer is more susceptible to erosion. Lightweight particles can be dislodged and carried long distances. Heavy particles like sand are likely to be moved only short distances. Loose or dry particles are also dislodged more easily.

Control of erosion

Erosion is generally caused by the effects of wind and water. It follows that erosion control methods are generally aimed at modifying these effects. Some of the most common control methods include:

- Prevention of erosion in the first place by careful land management
- Prevention of soil detachment by the use of cover materials such as trees, mulches, stubbles, matting and crops
- Crop production techniques (eg fertilising) to promote plant growth and hence surface cover
- Strip cropping (strips of cereal alternated with strips of pasture or other crop), hence no huge expanse is bare at any time
- Ploughing to destroy rills and contour planting to create small dams across a field, to retard or impound water flow
- Filling small gullies with mechanical equipment or converting the gully into a protected or grassed waterway
- Terracing of slopes to reduce the rate of runoff
- Conservation tillage
- Armouring of channels with rocks, tyres, concrete, timber, etc to prevent bank erosion
- Ploughing into clod sizes too big to be eroded, or ploughing into ridges
- Avoiding long periods of fallow
- Working organic matter into the soil
- Establishing windbreaks to modify wind action (see Chapter 9).

Salinity

Soil salts are compounds such as sodium chloride (common salt), sodium sulphate and various magnesium compounds in a crystal form. High salt levels in soils reduce the ability of plants to grow or even to survive. This may be caused by natural processes, but much occurs as a consequence of human action. Salinity has been described as the 'AIDS of the earth' and its influence is spreading throughout society; particularly in rural communities, where it has seriously affected crop production and caused economic hardship. Salinity problems have been grouped into two main types.

Dryland salinity is caused by the discharge of saline groundwater, where it intersects the surface topography. This often occurs at the base of hills or in depressions within hills or mountains. The large-scale clearing of forests since European settlement has seen increased 'recharge' of aquifers (where groundwater gathers in the ground) due to reduced evapotranspiration back to the atmosphere. The result has been a rise in groundwater levels, causing greater discharges to the surface.

Wetland salinity occurs where irrigation practices have caused a rise in the watertable, bringing saline groundwater within reach of plant roots. This is common on lower slopes and plains and is particularly common on riverine plains. The wetland salinity problem is compounded by rises in groundwater flow due to dryland salinisation processes higher in the catchment.

The presence of salinity is often indicated by:

- yellowing of pasture (NB: waterlogging or drought can cause this effect also)
- decline in crop plants, including yellowing, or burning of the tips of foliage
- bare patches developing
- reddish leaves on plantains
- the appearance of certain weeds and grasses (known as indicator plants)
- salt deposits appearing at the soil surface (in more severe cases)

Control methods for salinity

Many of the control methods for salinity are very expensive and require a strong commitment from governments if they are to be undertaken. They also require regional community cooperation because salinity does not recognise artificial boundaries. One of the major problems with salinity is that the area in which symptoms are most evident may be a fair distance from the cause of the problem. Thus we have saline groundwater discharging on the plains as a consequence of forest clearing high in adjacent hills where salinity problems may not be apparent. Many hill farmers are loath to change their practices for the sake of someone far away, especially if they have to suffer some economic loss as a result (eg the cost of tree planting and the loss of cropping area).

Some of the main methods of controlling salinity are:

- Pumping to lower groundwater levels, with the groundwater being pumped to evaporation basins or drainage systems.
- Careful irrigation practices to prevent a rise in, or to reduce, groundwater levels.
- 'Laser' grading to remove depressions and make best use of water on crop and grazing land.
- Revegetation of recharge areas and discharge sites.
- Engineering methods designed to remove saline water from crop land.
- Leaching suitable soils (eg raised crop beds).
- Identifying high recharge areas, which are often unproductive ridges and planting them with trees.
- Lucerne and phalaris, when substituted for other pasture species, have proven useful for lowering groundwater levels in some areas.
- Applying gypsum to surface soils with high sodium chloride levels; the calcium and magnesium in the gypsum will displace the sodium ions from soil particles, allowing them to be more readily leached by heavy irrigation or flood irrigation.
- Use of salt-tolerant species (eg wheatgrass)

Long-term solutions must involve treating the cause – not just the symptoms. This requires reducing recharge by appropriate plant cover in the catchment area, and more efficient irrigation practices. As less water enters the natural drainage system, the watertable starts to lower and return to levels closer to pre-clearing days.

Soil sodicity

Saline soils typically have a buildup of sodium chloride. In sodic soils, much of the chlorine has been washed away, leaving behind sodium ions (sodium atoms with a positive charge) which are attached to tiny clay particles in the soil. This makes the clay particles less able to stick together when wet – leading to unstable soils which may erode or become impermeable to both water and roots.

Affected soils erode easily and, in arid regions, sodic soils are susceptible to dust storms. In sloping areas, water easily removes the topsoil. Where the subsoil is sodic, water flowing below ground level can form tunnels which later collapse into gullies. The biggest problems occur when the top 5 cm of soil are sodic. However, when lower soil layers are affected it can also be a problem, as drainage is affected.

Sodic soils are usually treated with calcium-containing substances such as gypsum. Other ameliorants such as sulphur, aluminium and iron sulphates or iron pyrite can be effective. Gypsum is the most cost effective treatment that is readily available for treating large areas. In some cases, soils need to be deep ripped to allow penetration.

Soil structural decline

This causes a reduction in soil pore space, reducing the rate at which water can infiltrate and drain through the soil. It also reduces the available space for oxygen in the plant root zones, and makes it difficult for plant roots to penetrate through the soil. Some of the major consequences of soil structural decline are poor drainage, poor aeration, and hard pan surfaces which result in poor infiltration rates, and thus increased surface runoff and erosion.

Loss of soil structure is commonly caused by compaction due to human use of the soil (ie foot traffic on lawn areas, or repeated passage of machinery in crop areas). Poor cultivation techniques (eg cultivating wet soils and overcultivating) lead to a breakdown of soil structure and this also increases the likelihood of compaction. Soil structural decline can be prevented by farming practices that minimise cultivation and the passage of machinery. These include conservation tilling, selection of crops that require reduced cultivation and use of machinery at times less likely to cause compaction (ie when soils aren't too wet or when some protective covering vegetation may be present). For heavily compacted soils deep ripping may be necessary.

Soil acidification

This is a problem becoming increasingly common in cultivated soils. Soil acidification is the increase in the ratio of hydrogen ions in comparison to 'basic' ions within the soil. This ratio is expressed as pH, on a scale of 0–14 with 7 being neutral, below 7 acid, and above 7 alkaline. The pH of a soil can have major effects on plant growth, as various nutrients become unavailable for plant use at different pH levels. Most plants prefer a slightly acid soil, however an increase in soil acidity to the levels being found in many areas of cultivated land in Australia renders that land unsuitable for many crops, or requires extensive amelioration works to be undertaken.

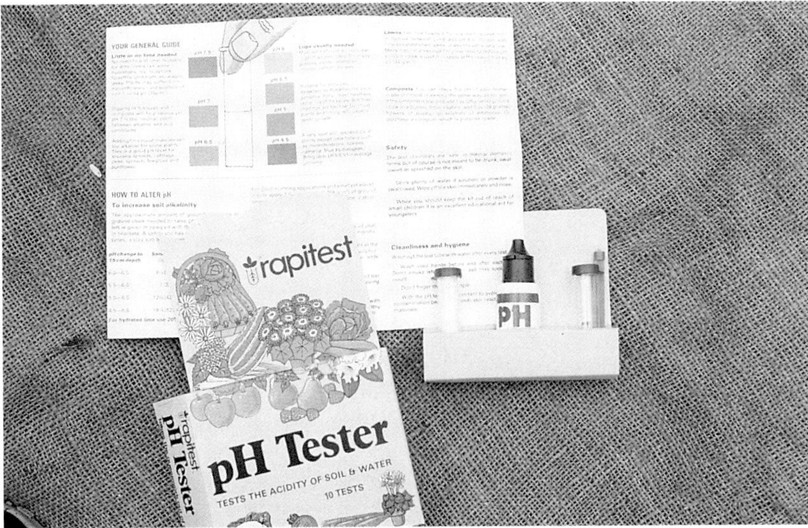

Figure 3.1 Soil pH test kit. Soil pH directly affects the growth of plants by limiting the availability of nutrients.

Causes of soil acidification

Acid soils can occur naturally but a number of agricultural practices have expanded the areas of such soils. The main causal factor is the growth of plants that use large amounts of basic ions (eg legumes); particularly when fertilisers that leave acidic residues (such as superphosphate or sulphate of ammonia) are used. Soil acidity is generally controlled by the addition of lime to the soil, by careful selection of fertiliser types and sometimes by changing crop types (see later section on adding soil ameliorants).

Build up of chemical residues

Although not as large a problem as some of the other types of soil degradation, the presence of chemical residues can be quite a problem on a local scale. These residues derive almost entirely from long-term accumulation after repeated use of pesticides, or use of pesticides or other chemicals with long residual effects. Some problems that result from chemical residues include toxic effects on crop species and contamination of workers, livestock and adjacent streams. Control is often difficult and may involve allowing contaminated areas to lie fallow, leaching affected areas, trying to deactivate or neutralise the chemicals, removing the contaminated soil, or selecting tolerant crops.

Improving soils

Nearly any soil can be 'improved' in some way to make it more suitable for agricultural or horticultural production. This is more readily done for small areas where the inputs required (eg soil additives, time, labour and machinery) are small, but can require considerable expenditure for large areas. Long-term increases in production will generally make such efforts very worthwhile. The following are common ways of improving soils using sustainable methods.

Adding organic matter

Most soils will benefit from the addition of organic matter, except those rare soils that are already high in organic matter such as peaty soils. Soils with good levels of organic matter are generally easily worked (ie they have a good 'tilth'). If you squeeze a handful of soil into a ball in your hand and it remains in a hard lump, then it has a poor tilth and hard clods will result when it is ploughed. If it crumbles, then it is well granulated – organic matter promotes granulation. Cultivated soils with good tilth are less subject to wind and water erosion.

Organic matter will improve the soil by:

- Helping to improve soil structure; this will also improve water penetration and drainage, as well as improving aeration. Adding organic matter is particularly valuable for poorly structured clay soils.
- Adding valuable nutrients to the soil
- Helping to retain moisture in well drained soils, eg sandy soil; every percentage point of soil organic matter is considered capable of holding the equivalent of 25 mm of rainfall.
- Acting as a buffer against sudden temperature or chemical changes which may affect plant growth
- Encouraging the activity of beneficial soil organisms such as earthworms

Figure 3.2 Processed manure is an alternative to artificial fertilisers in small-scale agricultural operations.

It may be difficult to increase the percentage of organic matter in a soil, but it is important to try to maintain that percentage. The average mineral soil contains around 2–5% organic matter. Organic content will drop if you remove plants from a soil and don't return organic material to the soil. Organic matter can be added in the following ways:

- Cultivate the roots of crop plants back into the soil when the plant has finished growing
- Add compost regularly (see section on compost in this chapter).
- Apply organic mulches regularly to the surface of the soil (see section on mulches in this chapter)
- Feed plants with manure (preferably well rotted) and other organic fertilisers
- Rotate crops to support organic soil content, eg use 70% of a farm for cash crops and grow a cover crop on the other 30%. The cover crop is then ploughed in, replenishing the lost organic content from the previous season (see the section on cover crops in Chapter 8).

Problems with organic materials

Soils containing high levels of materials such as peat, bark and sawdust can be very hard to re-wet if they are allowed to dry out. Organic materials can coat soil particles (particularly in sandy soils) and make the whole soil difficult to re-wet. In very bad cases water droplets sit on the soil surface and do not infiltrate into the soil at all. To cope with this problem you can do several things. If the soil is mechanically mixed it will assist wetting, then you can help keep it wet with mulching and frequent watering.

Phytotoxicity is where 'poisonous' parts of organic matter cause harm or even death to living plants. Phytotoxins can come from residue of decomposing micro-organisms, fresh plant residues dug into the soil, and even from the plants themselves. The older the plant is when it is incorporated into the soil, the more likely it is to be toxic, so avoid planting in that area for a while. Young green crops generally have only a low level of toxicity when incorporated into soil. These problems can be avoided to a degree if the residues are not dug in, but left on the surface as mulch or composted. Common phytotoxicity problems can also occur with mulches of fresh shredded or chipped pinebark or Eucalyptus. These materials should be composted for six to eight weeks before being used.

Decomposing fresh organic material releases carbon dioxide which can damage roots. Fresh organic materials (particularly animal wastes) may also release levels of ammonia gas that can cause burning to plant roots and foliage. Such wastes should be composted for a few weeks prior to being used, or used only in small amounts at a time. It is important not to put anything too fresh on plants and, wherever possible, to compost organic materials prior to use.

In warm, wet climates, the organic content of the soil can be low (under 0.5%), because organic material breaks down faster in these areas. This is particularly a problem in sandy soils. In these areas, add at least 4% compost (or organic material) to crop areas (eg vegetable beds) when you prepare them, and top up annually with the same amount.

Adding non-organic materials to soil

Light sandy soils can often be improved by the addition of materials with fine particles such as clay and silt. This will help improve moisture and nutrient retention. The addition of coarse materials such as sand to heavy clay soils can help improve drainage and water penetration. Generally, fairly high amounts need to be added to be effective. Any added material should be thoroughly mixed in.

Be careful to avoid adding material that may be contaminated in some way (eg with lots of weed seeds, pollutants, pests or diseases, or salts). This type of soil mixing is generally impractical for broadacre farming, but can be useful for improving soils on a smaller scale (eg intensive cropping systems).

Adding lime

This is the main way to raise the soil pH if it is too acid. Soils can be naturally acid, or may become too acidic when fertilisers such as sulphate of ammonia have been extensively used, where excessive manures or mulches are applied, or if plants that deplete the soil of calcium (eg legumes) have been grown. Lime might also be used if you are growing lime-loving plants such as cabbage, cauliflower and broccoli.

The main liming materials are:

- Crushed limestone (calcium carbonate) – the most commonly used and least expensive form of lime
- Dolomite – a mixture of calcium and magnesium carbonates is commonly used, especially in the nursery industry where soilless growing mixes are often used
- Quicklime – calcium oxide and builders' lime, also known as 'Limil'. More concentrated and expensive than limestone or dolomite, and can be easily over-used, raising the pH to a much higher level than desired. If a quick result is needed for small areas, then builders' lime is quite useful.

The amount of lime to be applied will depend on a number of factors:

- How acid the soil is
- The buffering capacity of the soil – or how resistant it is to a change of pH
- How acid the subsoil is
- The quality or purity of the liming material to be used (ie how much calcium carbonate it contains)
- How often the lime is to be applied
- What you want to grow (each plant having its own preferred pH range)

Table 2 Approximate amounts of calcium carbonate needed to raise the pH by one point of the top 10 cm of soils of different texture in grams per square metre (g/m2) of soil surface

Soil Texture	pH 4.5 – >5.5	pH 5.5 – >6.5
Sand, loamy sand	85	110
Sandy loam	130	195
Loam	195	240
Silty loam	280	320
Clay loam	320	410
Organic loam	680	790

Source: *Soil acidity and liming*, by R.W. Pearson and F. Adams (eds), Agronomy Series No 12, American Soc. Agronomy 1967

The percentage of calcium carbonate in the liming material used will generally be stated on the packaging or, for large lots, provided by the supplier.

Adding acidic materials to lower soil pH

Sometimes it is necessary to lower the soil pH to provide the ideal growing conditions for particular plants. To try and alter soils with a higher pH than 7.5 can become quite expensive and it is often best to simply grow plants that suit the alkaline conditions, or to slightly reduce the pH, rather than to try for major reductions in pH. This can be achieved on a large scale by the use of acidifying fertilisers, such as sulphate of ammonia and superphosphate, or by the regular additions of organic matter, in particular manures. These will generally take several years to be effective.

On a much smaller scale try the following:

- The addition of sulphur. Sulphur is oxidised into sulphuric acid by soil micro-organisms. This acid reacts with calcium carbonate in the soil to form gypsum, which has a pH close to neutral. The conversion of the alkaline calcium carbonate to gypsum therefore reduces soil pH. For soils that are neutral to slightly alkaline use between 25 grams for sands, to 100 grams for clays, of sulphur per m^2 to lower the pH in the top 10 cm of soil to around pH 6.0–6.5. This is equivalent to 250 kg of sulphur per hectare for sands, and 1000 kg of sulphur per hectare for clays. To achieve greater reductions would necessitate quite extensive applications of sulphur, which would be very expensive. For quickest results mix the sulphur into the soil rather than spreading it on the soil surface.

- Adding material such as peat or coconut fibre, which has considerable acidifying abilities. One cubic metre of peat has an equivalent acidifying effect to about 320–640 grams of sulphur. To lower the pH one point in the top 10 cm of soil, one cubic metre of peat incorporated into the soil will be effective over an area of about 3.25 m^2 for clay soils, ranging up to about 13 m^2 for sandy soils.
- Ferrous sulphate can be used at a rate of around 50–150 grams per m^2. Diluted solutions of iron sulphate or phosphoric acid can also be used.

Adding gypsum

Gypsum is commonly applied to hard packed or poorly structured clay soils. It has the ability to cause clay particles to aggregate together in small crumbs (or peds), thereby improving structure. It is also used to reclaim sodic soils. Gypsum contains around 23–25% calcium and about 15% sulphur. It will not affect soil pH to any great extent. Rates of up to two tonnes per hectare are used to treat hard-setting cereal growing soils, and up to 10 tonne per hectare to reclaim saline-sodic clay soils.

NB: The previous three treatments require moist soil conditions over several months to have a noticeable effect. It is important not to expect immediate results.

Cultivation techniques

Cultivation involves ripping, digging, scratching or mixing the soil. This may be done for any of a range of reasons, including:

- To mix in compost, fertiliser or a cover crop
- To kill weeds
- To break an impermeable layer on the surface to allow water or nutrients to penetrate
- To improve drainage
- To allow for better plant root penetration
- To break up an impermeable subsurface layer

Cultivation can, however, also cause problems. Overcultivation or regular turning can damage soil structure. Subsequent cultivation damages some of the small aggregates, allowing the organic matter which binds these aggregates to be consumed by micro-organisms. Cultivation is one of the main factors in causing erosion and soil structure decline. It can also change soil drainage patterns and can cause the fertile top layer to be diminished by mixing it up with lower soil layers.

Minimal cultivation is normally preferred in sustainable agriculture but cultivation is a necessary part of any farming operation. Some ways to minimise damage include:

- Tilling only where necessary, eg leaving strips of unturned land staggered with tilled soil.
- Not tilling when soil is overly wet. A simple test is to take a handful of soil and squeeze it in your hand. Moisture levels should be no more than you would get by squeezing a sponge dry.
- It is preferable to use discs or ploughs rather than rotary hoes or tillers which mix the soil more.

Conservation tillage

This aims to reduce tillage operations or cultivations to only one or two passes per crop. It has been made possible by the use of herbicides to kill crop residues or pasture prior to planting, and the development of direct-drilling seeding machinery capable of seeding through stubble. For some farmers the extensive use of these herbicides does not fit in with their view of what sustainable farming should be, however for many farmers the disadvantages of using such herbicides are more than offset by the benefits of maintaining or improving soil characteristics, in particular structure. Conservation tillage has been shown to give sustained, improved yields when compared with cultivated paddocks. There are also considerable benefits in reduced labour costs, less wear and tear on equipment, and decreased fuel costs, as a result of the reduced number of passes required.

Stubble retention (from the previous crop) is a major component of conservation tillage. The stubble provides a protective layer on the soil, reducing evaporation losses, and reducing the impact of rain drops. This prevents the formation of surface crusts, and improves aeration and water infiltration. There is also a reduction in diseases of legume crops that are spread by raindrop splash. Soil micro organisms have also been shown to increase in numbers, further helping to improve soil structure and fertility.

The biggest barrier to the use of conservation tilling has been the cost of buying or modifying tillage and seeding machinery. Conventional seeding machinery has had difficulty coping with the retained stubble. As this method of cultivation has increased in popularity, there has been extensive development of new machinery that can cope with such demands. The gains, however, are seen to more than outweigh the cost outlays, and this method of farming is sure to increase.

Source: *There's no money in dust: A guide for farmers modifying their seeders for conservation tillage* by Nicholas Bate. Published by Farm Advance, Available from DPI (Vic), P.O. Box 3100, Bendigo DC, Vic 3554, Australia.

Plant nutrition

In order to survive, plants must be supplied with appropriate amounts of certain nutrients. About 16 nutrients are needed by all plants, with three of these, carbon (C), hydrogen (H) and oxygen (O) being obtained by the plant from air, water and oxygen. The rest of these nutrients are naturally obtained from the soil and these are sometimes called the mineral nutrients. Some plants also require various other nutrients to thrive.

Plant nutrients are divided into two groups, the major (or macro nutrients) and trace (or micronutrients) elements. The major elements are carbon (C), hydrogen (H), oxygen (O), nitrogen (N), phosphorus (P), potassium (K), calcium (C), magnesium (Mg), and sulphur (S).

The mineral nutrients used in the largest amounts are nitrogen, phosphorus and potassium. These are the most frequently applied as fertilisers, either from organic or inorganic sources, to encourage plant growth.

Every nutrient has its purpose, and a deficiency or oversupply of even a minor nutrient can have a major effect on the plant. Deficiencies can be difficult to detect, but as time passes symptoms will appear. Signs are stunted growth, unhealthy leaves that may be mottled, stunted and dying off, distorted stems and undeveloped root systems. If a nutrient is easily dissolved, the older leaves will be affected first, otherwise the growing tips, ie the new leaves, will be affected.

An oversupply of nutrients may initially cause extra growth, but may then become toxic, and plant growth will be reduced. Deficiencies are not always a result of the nutrient being absent: it may be the nutrient is being held in some form which prevents the plant taking it up, for example it could be attached to an insoluble material (known as 'immobilisation' or being 'locked up'), or be affected by pH.

In simple terms, in order to ensure healthy plant growth, do not let plants suffer from nutrient deficiency or toxicity. For organic farmers and gardeners, supplying minor trace elements to suffering plants may seem a bit daunting. Provided you use a wide range of organic fertilisers as sources of organic matter, it is unlikely that plants will suffer from any deficiencies. In fact, soils high in organic matter hold more nutrients than inorganic soils.

Nutrients can also be readily lost from the soil through erosion, through leaching (the loss of soluble nutrients down through the soil profile in soil water), through conversion of nutrients to gaseous forms (eg ammonia gas which escapes to the atmosphere), and through the removal of plant material (eg crops). Compensation for the loss of nutrients by such means should be a major priority of the sustainable farmer. Chemical fertilisers will compensate for these losses in the short term; however the sustainable farmer should be looking more towards cover crops, mineral powders and composts that release nutrients slowly. This requires careful planning as benefits will only accumulate gradually.

Soil pH

Soil pH is measured on a scale of 1–14. A pH of 7 is called neutral, below this the soil is called acid and above, alkaline (or limey). Plants require a pH between 4.5 and 9 – most prefer a pH between 5 and 7. The main effect of pH is on the availability of nutrients. Plants prefer different pH levels due to their roots' ability to extract different nutrients. If a plant is growing in a pH it is not adapted to, it may either be unable to extract enough nutrients or may extract so much that it becomes oversupplied. For example, plants grown in alkaline soils may suffer from iron deficiencies.

Choosing the right fertiliser

Using the right fertiliser helps to minimise wastage, reduce costs, and reduce negative effects on the environment, while maximising plant growth.

- Timing is important so as not to waste fertiliser. In winter, some plants may be dormant so the fertiliser will not be taken up. Heavy feeding at the wrong time of year can also cause fruit trees to produce plenty of leaves at the expense of fruit.

- Commercial fertilisers are available for certain types of plants (eg Citrus Food, Rose Food) or as general preparations to suit most plants. However, some are produced from non-renewable resources.
- Quick-release or soluble fertilisers are very mobile, which makes them easier for the plants to access, but unfortunately most of the nutrients can be leached into streams or ground water, eventually ending up in rivers, bays, dams and estuaries, causing problems such as algal blooms.
- Using slow-release fertiliser can be a more efficient way of feeding plants, but again these may not be made from renewable materials.
- Home-made fertilisers can be prepared using compost, animal manures and mulch material. Some plants themselves are excellent sources of nutrients, including legumes (eg lucerne). Often weeds are able to absorb minor nutrients from the soil, so they can also be used, if care is taken to ensure that the weeds have not set seed and will not re-establish. A handy way to make your own liquid fertiliser is to quarter fill a large container with weeds, and/or manures and legumes, top up with water to cover the material, and leave it all to stew for a couple of weeks, stirring occasionally. The resulting dark liquid should be diluted with water (1:100) and applied to the soil or used as a foliar (to the leaves) application. The brew may be regularly topped up with water and other ingredients. Be careful to keep it covered or agitate it regularly to prevent mosquitoes breeding in the container.

The golden rules for using any liquid manures are 'diluted' and 'frequent'. There comes a point where a strong organic manure can be as disastrous as chemical manures used injudiciously. Urine is an excellent liquid manure if diluted about 1:20 but, if used at full strength, it will kill almost any plant.

Table 3 Symptoms of nutrient deficiencies

Nutrient Deficient	Symptoms
Nitrogen (N)	Leaves turning pale green, then yellow; in some cases the leaves have red or purple tinting; oldest leaves dry to a light brown colour; leaves are small; stalks are stunted
Phosphorus (P)	General stunting, little branching, so plants look spindly, leaves take on blue grey and purple colours before gradually yellowing
Potassium (K)	Leaves first become dull grey-green; leaf margins and tips first start to turn yellow in spots and yellowing expands with the firstly infected areas 'scorching' and dying; stalks are thin and shortened
Magnesium (Mg)	Leaves look mottled, with patchy yellowing, sometimes accompanied by reddening, starting around the edges; tips and edges are cupped; sometimes brilliant colouring, especially around the margins; light tan, dead spots appear in the yellow areas, expanding until leaves die

Natural fertilisers

Nutrients can be added to soil by digging in kitchen scraps, animal manures, cover crops, natural minerals such as rock dusts, and synthetic chemical fertilisers. Nutrients are also obtained from irrigation water, rainfall, from the atmosphere (ie micro organisms converting atmospheric nitrogen) and from the natural weathering of rock and soil itself.

The source is unimportant to the plant; nitrogen from animal manure is exactly the same as nitrogen from sulphate of ammonia, and phosphorus from rock dusts is exactly the same as phosphorus from superphosphate. The choice of which source of nutrients to use should depend on the effect that it will have on the soil.

Artificial fertilisers are easier to apply and manage than animal manures and organic fertilisers, but can create major soil problems, in particular soil acidification. These fertilisers release nutrients quickly so nutrients are easily washed through the soil where they can pollute rivers and creeks. Organic fertilisers generally don't cause these problems, and have the added advantage of improving soil structure, and promoting beneficial soil life. As such, they represent a favourable option for proponents of sustainable agriculture.

Table 4 Types of natural fertilisers to use and their average relative nutrient content, rate and availability

Fertiliser	%Nitrogen	%Phosphorus	%Potassium availability	Nutrient	pH
Sewage sludge	3.0	2.0	0.3	Slow	Acid
Cocoa shell meal	2.5	1.5	1.5	Slow	Neutral
Peat	2.0	0.3	0.7	Very slow	Acid
Fish meal	10.0	4.0	0.0	Slow	Acid
Blood meal	12.0	1.5	0.8	Slow	Acid
Hoof and horn	12.0	2.0	0.0	Slow	Neutral
Bone meal	3.5	2.0	0.0	Slow	Alkaline
Powdered rock					
Phosphate	0.0	33.0	0.0	Very slow	Alkaline
Seaweed	1.0	0.0	5.0	Slow	Acid
Wood ash	0.0	2.0	4.0	Slow Very	Alkaline
Urea	45.0	0.0	0.0	Fast	Very acid
Manure (fresh):					
Cow	0.25	0.15	0.25	Medium	Acid
Horse	0.3	15.0	0.5	Medium	Acid
Sheep	0.6	0.35	0.75	Medium	Acid
Poultry	2-6	1-4	0.5-3	Fast	Acid
Pig	0.3	0.3	0.3	Medium	Acid
Green bracken leaves	1.5	0.2	1.5	-	-
Apple cores	0.5	0.02	0.1	-	-
Coffee grounds (dried)	2.0	0.4	0.7	-	-
Citrus peel	0.2	0.1	0.2	-	-
Peanut shells	0.8	0.15	0.5	-	-
Tea leaves	4.2	0.6	0.4	-	-

Source: *Fertility Gardening* by Hills (1981), published by David and Charles; *Hydro Story* by Brenizer (1977), published by Nolo Press

A look at organic fertilisers

It is important that all of the materials used in either composting or as fertilisers be free of chemical contamination. This includes antibiotics, growth stimulants and other products that may have been fed to livestock.

Animal manures

All animal manures are useful as fertilisers. When mixed into compost as part of the composting process, the final material provides excellent all-purpose fertiliser.

Manures can be used directly on plants; however they vary greatly in their nutrient content. It is impossible to give accurate figures on the micronutrients in animal manures, because they can vary so much, depending on what the animal eats.

Having some idea of the nitrogen content is important in any manure you use. High levels of nitrogen can burn the plant roots. Well-rotted cow, sheep, horse and goat manures are generally safe (nitrogen is not too strong).

Pulverised and partly composted cow manure can be used generously on a bed in preparation to plant seedlings on the same day, provided it is thoroughly mixed in the top 6–8 cm of soil. (The same treatment with an equal amount of poultry or pigeon manure would result in disaster, with most, if not all of the seedlings dying within a few days.)

Poultry manure

Poultry manure may be available either as pure manure without litter, or as dry deep litter material (mixed together with woodshavings). When using pure poultry manure, make sure that the chicken farmers have not sprayed the manure heaps with insecticides, a common practice in some farms. Deep litter poultry manure is always a safer bet because it is far less likely to have been sprayed (chickens scratching the litter eat insects and spraying is unnecessary).

Blood and bone meal

Dried blood used in a number of commercial organic manures has probably the highest nitrogen content of any organic manure – about 11.5%. It is, however, expensive. Blood meal has approximately 1% phosphorus. Bone meal contains around 4% nitrogen but over 20% phosphorus.

Rock dusts

Rock dusts are simply ground up or crushed rocks. Natural weathering in farms or gardens gradually leaches out many of the original nutrient reserves in the soil. Many of these nutrients originally came from silt or weathered rock, so it is argued by some organic gardeners that applying rock dust may replenish the nutrients in the soil. Some experts are still sceptical about the benefits of rock dusts. Some rock dusts used in agriculture and horticulture are:

- Gypsum to supply calcium and sulphur
- Dolomite to supply calcium and magnesium
- Limestone to supply calcium
- Scoria to supply iron
- Basalt to supply calcium, magnesium, phosphorus, potassium and a range of minor nutrients

Seaweed

- Many types of seaweed do not contain cellulose like other plants. This causes it to rot down faster than other plants.
- It is ideal to use in a compost heap
- Wash harmful surface salt off the plants before use

- It is a rich source of boron, iodine, calcium, magnesium, sodium and other trace elements
- Readily available nutrients can cause nutrient toxicities (causing leaf burn) if overused
- It is safe to mulch plants with a generous layer of seaweed about once every four years, making it suitable for long-term crops such as fruit and vine plants
- Different types of seaweed have different nutrient values. The kelps (the large flat, brown strap-like seaweeds) are the best source of potash.

Seaweed extracts

There is a range of liquid fertilisers on the market which vary in quality. Although seaweed doesn't have high levels of nitrogen, phosphorus or potassium, it does contain a broad spectrum of micronutrients, and many people attest to strong plant growth, vigour and general good health when they use seaweed extracts (such as Maxicrop and Seasol). The fact that this can't be attributed only to the direct absorption of those nutrients supports the probability of other benefits (ie that plant nutrition is enhanced by increased bacterial, enzyme and other biological factors in the soil, that in turn provide far better nutritional status).

It is difficult to overdose plants with liquid seaweed. Some people even dip cuttings in it to encourage root formation.

Soil life

Earthworms

Earthworms are very important in the farm. A plentiful supply of earthworms in your soil is a good indication of a healthy soil. As earthworms work, they pass soil through their bodies, mixing layers of soil and leaving loosely packed material in their tracks. Along with micro-organisms, they also help to break down organic matter, turning it into humus which is an important soil conditioner.

Earthworms thrive on organic matter. Organic matter sometimes lacks the nitrogen earthworms need. If worm numbers are low, it may be useful to add manure or some other high nitrogen fertiliser to encourage their growth and reproduction. Nitrogen-fixing plants such as legumes also attract and support earthworms. Commercially grown earthworms can be added to soils and compost heaps, as long as they have a high organic matter content. The most commonly available worm varieties are *Lumbricus rubellus* (red worm) and *Helodrilus foetidus* (tiger worm).

Mycorrhiza

Various fungi are often found to be associated with particular types of plants. These fungi are found to have a special relationship with tree roots forming a structure called mycorrhiza. Most healthy trees will tend to show this condition. The fungus appears to get nutrition from the tree, whilst not harming the tree itself. The presence of the fungi can assist in the tree's growth by increasing the tree root's absorptive area. Eventually infected roots are shed by the tree and the fungus utilises them as food.

Many mycorrhiza live in a symbiotic (mutually dependent) relationship to a plant, many of which are legumes such as peas, beans, lupins, and wind shelter trees like she-oaks (Casuarina).

Nitrogen fixing

Some plants, such as the legumes, have the ability to fix atmospheric nitrogen into the soil. What this means is that the plant is able to convert nitrogen in the air into a compound that can be used in the soil by plants. This is carried out by micro-organisms, such as Rhizobium bacteria, that live in swellings on the roots of legume plants. These plants can be a valuable source of nitrogen to a soil.

Composting

Where the opportunity arises, a farmer should make and use compost to improve farm soils. Any organic material will usually be suitable for compost (however be cautious of material that may contain dangerous chemicals – manure/shavings mix from a poultry farm may often contain a cocktail of chemicals used with battery hens, which you can usually smell in the manure).

Cheap sources of compost are very valuable, particularly if soils are low in organic content. These may include woodshavings from timber production, spoilt hay/unwanted feed at the end of an exceptionally good season; or compostable factory wastes (eg rice hulls, poppy straw, sugar cane waste).

A simple method of composting can be to mound material (at least 1 m high after settling), leave it for six months in wet weather is wet and one year or more if the weather is dry. After this, grow a crop on the mound (eg pumpkins or potatoes) then spread the composted material over your paddocks as a mulch, or cultivate it in.

The composting process

Any organic material, if left long enough, will eventually rot down due to the action of micro organisms. Composting is simply a way to control this process by speeding up the rate of decomposition and minimising nutrient losses.

Compost incorporated into the soil on the farm, whether broadcasted over the land or actually ploughed in, will improve the physical and chemical features of the soil. This improvement, like most other sustainable practices, will not occur immediately. Time is required.

A farmer may have access to bulk supplies of organic materials. In this situation, it is recommended they accumulate fresh material and stockpile it until a suitable time has elapsed (about two to four months) before it is used on the farm. If material is well aged to begin with, then immediate use may be possible.

The raw material for any compost is organic matter. This may be in the form of unharvested plant material, windbreak prunings, grass clippings, dead animals and birds, manure, household or farm organic garbage, hay, straw, paper even sawdust. The smaller and finer the particles, the quicker the composting process will be.

Diseased plant material should not be used in compost as it may contaminate new areas when the compost is spread around at a later date. If the farmer has access to dead animal products such as bone, skins, offal or similar, it is important to consider health regulations. Such products also tend to attract vermin and insects to the compost site.

Animal manures are an excellent source of matter for compost. The most commonly used are sheep, cattle, poultry, horse and pig manures, although others can be quite valu-

Figure 3.3 A large compost heap.

able if you can obtain them in large enough quantities. Animal manures need to be composted for a minimum of six weeks to prevent problems such as burning of leaves and roots from the presence of high levels of ammonium ions in the fresh manure. The ammonium ions are rapidly lost during composting.

Composting of manures is also valuable in reducing potential weed problems that may arise due to the presence of large quantities of seed eaten by grazing animals. The seed passes through the animal and is deposited in the animal droppings where the nutrients present in the manure and the warmth generated as it decomposes create an ideal environment for the seed to germinate. Incorporating manure in a compost heap results in much higher temperatures that will kill a large percentage of the weed and grass seeds prior to, or just after, germination.

The basic conditions of compost the farmer needs to be aware of are:

- Moisture – should be between 40 and 60%. Take a handful of the composting material from 15–20 cm deep into the heap/mound of composting material, and squeeze it. It should be about as moist as a moderately squeezed wet sponge. If it is too dry, add water to the heap. If it is too wet you may need to cover the heap with plastic, or turn it over regularly to allow for more evaporation to occur.
- Oxygen – is incorporated by aerating and turning the heap over occasionally
- Temperature – should be between 40 and 60°C
- pH – will change during the various stages of decomposition; generally you need not do anything to alter pH
- C/N Ratio – The ratio of carbon: nitrogen should be in the range of 25–30:1.

If the compost process is permitted to fall outside these guidelines, then the compost will take longer to produce and may lose some nutritional value.

What is the C/N ratio?

For effective composting to occur, the micro organisms that break down the plant materials require food in the form of nitrogen, phosphorous and potassium. Phosphorous and potassium are generally quite plentiful in composting materials but there is often a lack of nitrogen. The most important requirement is the ratio of the percent carbon (C) in the materials, to the percent nitrogen (N). This is called the Carbon/Nitrogen ratio. Raw garbage, for example, has 25 times as much carbon as it has nitrogen, so its C/N ratio is simply expressed as the number 25. A C/N ratio of around 30 is required for compost activity to take place at an optimum rate. To get a suitable C/N ratio it is necessary to mix materials with a high C/N ratio such as sawdust with materials that have a low C/N ratio such as manures.

Table 5 C/N Ratios Of Some Compostable Materials

Material	C/N Ratio
Ash leaves	30
Blood meal	4
Bracken leaves	48
Cabbage heads	12
Chicken litter (average – with sawdust)	10
Chicken manure (no sawdust)	7
Clover (old plants)	20–30
Clover (young seedlings)	12
Comfrey leaves	10
Composted pine bark (average)	200
Corn stalks, leaves and cobs	50–100
Cow manure	15
Eucalyptus bark	250
Eucalyptus sawdust (fresh)	500
Fruit wastes	35
Grass clippings	20–25
Green ryegrass	36
Leaves (mature)	60
Lucerne hay	13
Mature compost	10
Mixed weeds	19
Oak leaves	50
Paper	170
Pea or bean plants	15
Peanut shells	2
Pinebark (fresh)	500
Poultry litter	10–11
Poultry manure	7
Rice hulls	140
Sawdust (old)	200
Seaweed (average)	20–25
Straw (general)	100
Straw (Oat)	48
Straw (Wheat)	128
Tomato leaves and stems	12
Vegetable peelings	20 to 30
Well rotted manure (average)	20

The best type of compost will result from using the best type of organic material. The organic matter used should have a ratio of carbon to nitrogen atoms averaging between 25 and 30. Despite this consideration of what is ideal, you can use absolutely anything organic on your compost heap if you wish. Just keep in mind that if the carbon/nitrogen is unsuitable, it may take a very long time for decomposition to occur.

What can go wrong?

The main reasons for compost systems failing are as follows:

- They get too WET; if a foul odour is present, this is probably the case; extra turning or adding dry materials can overcome this problem
- They get too DRY; if the centre of the heap is dusty, this is far too dry
- Lack of NUTRIENTS such as phosphorus or potassium may reduce the rate of decomposition; organic materials high in these nutrients can be added to the compost heap to rectify this problem
- Carbon/nitrogen ratio is incorrect; lack of nitrogen because of too much high ratio material is common
- Presence of rats or feral activity; use of raw meat and protein products can result in putrid smells and populations of rats and other scavengers; avoid the use of such products.

The finished product

When compost is ready to use:

- It is crumbly and generally an even texture (material such as straw or flower stems might still be intact)
- It should drain well, but still have good moisture-holding capacity
- It should be dark in colour
- It should smell earthy, not rotten or mouldy
- The high temperatures that occurred in the centre of the heap during decomposition should have dropped
- There should be few, if any, disease organisms or weeds left alive

Guidelines for using compost

- Compost can be used either as a mulch, spread on the surface of the ground, or dug in (mixed with soil), to improve the structure of soil
- In temperate areas, the best time to add compost is in autumn; let it lie on the surface over winter then dig it in spring.
- Do not leave compost too long (particularly in warm weather) before using it, as nutrients can be lost over time
- Don't plant in pure compost alone. Compost is good for most plants, but doesn't have everything a plant needs; soil is necessary too

Sheet composting

This method is useful for farmers. The area may have to be lightly cultivated in order to sow a green manure crop such as soybeans, clover or cowpeas. After the green manure has germinated and before the nitrogen-rich plants reach maturity, compost materials are spread over the area. Low nitrogen materials such as sawdust, corncobs and wood chips can be spread without any fear of causing nitrogen shortages later on. After spreading, the whole mass is worked into the soil, preferably with a rotary hoe. The aim should be to incorporate the organic material evenly into the top 10 cm of soil. It is also a good idea to add limestone, phosphate rock, granite dust, or other natural mineral fertilisers along with the other sheet compost ingredients, since the decay of the organic matter will assist with the release of the nutrients locked up in those relatively insoluble fertilisers.

Worm digester or earth worm farm method

Many home gardeners are now getting into worm farming, and the principles can also be used in the farm situation. When earthworms breed, they work with raw materials and turn them into rich fine compost. Given the right conditions and materials it is possible to produce compost in 60 days. In addition, the compost is rich in castings or manure of the earthworms which is superior to animal manures.

Composting with earthworms is usually done in wooden cases of about 1 metre square and about 0.5 metre high. A mixture of raw materials is placed in the boxes, typically containing about 70% weeds, leaves, grass clippings etc; about 15% manure and 12% topsoil. If no manure is available, kitchen wastes can be substituted. All the ingredients are thoroughly mixed and then placed in the boxes. It is advisable to use purchased worms as they thrive best under controlled conditions. Tiger worms and red wrigglers are two of the best varieties. Before placing the worms into the boxes, check that the heaps have not become too hot, or else the earthworms will either leave or perish. The material in the boxes should be piled to a height of about 0.5 metres and kept sufficiently watered. Care should be taken not to add too much water otherwise air will be excluded from the heap. It is the combined action of the earthworms, bacteria and fungi that produces the best kind of compost.

The whole process takes about 60 days. Remove half a box of material and fill it with the raw materials. In 60 days the new material will be completely composted. It is advisable to feed the worms with something equivalent to chicken mash, but you can make your own feed using ground corn and coffee grounds.

Mulches

Mulching is a cultural practice commonly used by orchardists, some vine growers, and the home gardener. The broadacre farmer would have little use for mulching.

The principle requires the farmer to cover the ground under cultivation with a 'mulching material'. The extent of coverage may be limited to small zones around each plant or may cover the entire length of the rows.

Mulching will do the following:

- Reduce the need for watering (by preventing evaporation from the soil surface)
- Minimise temperature fluctuation in roots ie:
  Reduces frost damage
  Keeps roots cool in summer and warm in winter
- Control weed growth
- Reduce wind and water erosion
- Improve the appearance of a garden
- Organic mulches provide nutrients as they decompose.

Mulch may be classified as either organic or inorganic. Organic mulches include hay, straw, compost, carpet, and underfelt – in other words, anything that was organic at one stage. Inorganic mulches include stones, plastic sheeting and plastic weed mats.

Before laying any mulch it is very important to control any weeds in the area where the mulch is to be laid. Control options include:

- Lay wet newspaper thickly first then place mulch on top. This thick layer will cut out sunlight to weeds and should kill most troublesome weeds. This is only suitable for mulching small areas
- Forget the newspaper and just use very thick mulch (at least 20 cm thick)
- Cultivate the area to disturb the weeds and sever the plant from the roots; if this is done on a hot day and the weeds are left exposed to the sun, the weeds should be dead by the end of the day
- Spray the weeds with a killing solution then apply mulch

This last option raises many arguments for sustainable farmers. One of the safest weed control methods is by direct application of boiled water. This leaves no harmful residues in the ground, only good old H_2O. One method which causes the greatest debate is the use of glyphosate (ie Roundup, Zero, etc). This chemical reportedly has no, or very little, residual life and becomes neutralised in contact with organic matter.

General rules for mulching

- Most organic materials which haven't been composted will draw nitrogen from the soil. When used as a mulch they starve plants of nutrient, therefore you need to apply a side dressing of slow release nitrogen (such as blood and bone) around the base of plants (eg with straw, hay, shavings, leaf litter)
- A layer of newspaper or cardboard underneath reduces the thickness of mulch needed (saves money)
- Mulch should be shallower around the base of plant. Create a basin in the mulch around the stem otherwise mulch can cause collar rot; this is where the trunk at ground level starts to die off due to excess heat generated by the composting mulch and fungal infection
- Find out what mulching materials are available in your locality
- The type of mulch available and the cost varies from place to place
- Some light fluffy mulches settle to form a thinner layer (ie if you put it on 20 cm thick after a month it is likely to be only 10 cm thick). In this case you can renew mulches on a regular basis, place the mulch on extra thick to compensate for the shrinkage, or use another product.

Table 6 Types of mulches

Type	Cost	Availability	Comments
Woodshavings	Cheap	Sawmills, also look under sawdust in the telephone book	Avoid too much fine dust in the shavings, settles to thinner layer. Medium lasting, taking several years to decompose
Wood chip	Medium	Good in forestry areas and nearby cities	Can vary in quality and appearance from splinters to chunks; long lasting, semi permanent, very slow rate of decomposition
Straw	Medium	Better in rural areas	Often contaminated with grass seeds; only lasts one season, decomposing relatively fast
Lucerne hay	Medium	Better in rural areas	Unlikely to have weed seeds, looks good and works well. Usually only lasts one season, or two if laid on thick

Table 6 Types of mulches (continued)

Type	Cost	Availability	Comments
Seaweed	Not sold	Though generally not sold, it can be collected from the seaside (check with local authorities first)	Needs to be washed to remove salt, then it becomes a very good mulch, a good source of micronutrients.
Leaves	Not sold	Collected from below deciduous trees in autumn	Eucalypt and conifer leaves have toxins which harm some plants. Different types last for different periods of time (eg oak leaves don't rot down as fast as ash).
Lawn clippings	Not sold	Easy to collect when mowing or slashing	Decompose very fast. This can cause nitrogen deficiency if used in thick layers. Best used only 1 cm thick and topped up every few months as they rot
Manure	Cheap to medium	Good	Can burn if too thick, a short-term mulch, can bring weed seeds
Black plastic	Cheap	Very good	Only suitable on raised beds (eg as used with strawberries) Can sweat underneath becoming smelly. If top isn't washed clean by rain, it collects dust in which weeds grow. Stops rain wetting soil below – plants can become too dry under plastic. Can make the soil inactive! Generally it is not recommended, this is not an organic product.
Synthetic mulch fabrics	Medium to expensive	Readily available	Long lasting, their appearance isn't always what is desired; can be covered with a thin layer of bark for effect, not organic therefore not recommended
Paper	Cheap	Readily available	Used either shredded or as sheets; can be covered with bark for better effect; rots down in one to four years depending on thickness; do not use coloured newsprint as this contains heavy metals
Compost	Cheap to expensive	Varies	Some types (eg spent mushroom compost) can be in good supply and very cheap in some places. Well rotted compost is a good source of nutrients and a very good general mulch. It does not necessarily deter weeds. Some composts may be contaminated with weed seeds.
Peat	Expensive	Generally good	Very acidic, thus not good on lime loving plants; a non-renewable resource
Coconut fibre	Medium to expensive	Becoming more widely available	Similar properties to peat, but made from a renewable resource
Pine bark	Medium	Good in areas with large pine plantations	Can contain toxins if too fresh; should be well composted before use; larger chunky bark lasts longer
Tan bark	Expensive	Not generally available these days	Tanneries tend not to use bark as they did in the past

NB: The comments above are generalisations. Things do vary from place to place and some details might be quite different in different countries.

4

Water management

Water, its quantity and quality, can be a major determining factor in the success or failure of a farm. These features also have an influence on determining how the water will be used on the farm.

Water is commonly used on farms for:

- irrigating crops
- drinking (human and animal use)
- washing/sanitation
- aquaculture

Sources of water for farms might include direct collection of rain (into tanks), underground water (bores or springs), dams, lakes, creeks, river, atmosphere catching (condensation on the foliage of trees that drips to the ground), recycled waste water, desalination of sea water or, in some instances, connections to town water supplies.

Methods of water storage

Weir (watercourse dam)

In many places it is illegal to divert or stop the flow of a natural watercourse by damming; however, in such cases it may be permissible to build a weir to create a sump or to divert water into an off-stream storage dam or tank. Before doing so it is important that you contact the relevant water authority to discuss the legal aspects involved.

Hillside dam

The hillside dam, usually three-sided, is a cut and fill construction into the side of a prominent hillside. The embankment material is gouged from the hillside, forming a pocket-like effect. Water flows into this dam by sheet flow and diversion banks can be used to increase the amount of runoff collected.

Gully dam

This type of dam is created by building an earth wall across a natural drainage line between two ridges. The water is stored at a higher elevation than the surrounding grass flats, which can then be flood-irrigated by gravity. Underground pipes can be used to transport water to stock drinking troughs.

Tank

A tank designed to collect/store rainwater or bore water, usually made from concrete, galvanised iron or fibreglass.

Excavated tank

Below-ground level water catchment area usually restricted to flat ground.

Rainwater collection and storage

Few farms are connected to the mains water supply. Most farmers rely on rainwater collected and stored in tanks or dams, bore water pumped from underground streams or fresh water pumped from natural water courses.

Rainwater is collected from roofs and chanelled into storage tanks. When choosing a tank, consider:

- the roof catchment area – this determines how much water can be collected
- the tank size – the volume of water that can be stored
- your water requirements – for domestic, garden and farm use

To maintain water quality, ensure the tank excludes light as much as possible to prevent the growth of algae, and has effective inlet strainers and tight-fitting lids to prevent leaves, insects and other debris contaminating the water. A diverter trap or similar device can be installed to prevent accumulated debris being washed into the tank.

Regular maintenance includes keeping gutters clear of leaves and other debris, cleaning the inlet strainer and getting rid of mosquito larvae. A film of liquid paraffin will prevent the mosquitoes breeding – use 1 L of oil per 20 kL of tank capacity at the end of winter, and again in summer.

If the water is contaminated by bacteria, add non-stabilised chlorine such as calcium hypochlorite 60–70% or sodium hypochlorite 12.5%. The initial dosage will disinfect the tank, while weekly treatments may be required to maintain a safe water supply. Check with the chemical supplier for recommended dosages and application methods.

Bore water

Many farmers are able to access fresh groundwater stored in aquifers below the ground surface. Depending on the quality of the groundwater, it may be suitable for domestic, stock and irrigation – a complete water analysis should be carried out to determine the overall suitability of the water.

Drilling for water can be expensive – initial attempts often result in 'dry' holes (bores that yield no or insufficient water) and the process may need to be repeated several times or to a greater depth before reaching a satisfactory aquifer. Contact a local drilling contractor and/or a specialist water adviser for advice on the best location for the bore.

Some problems that may occur with bores:

- Decreased water supply – drought and excessive demands on the aquifer system will cause the water level to drop. Too many bores tapping into the same aquifer can deplete individual bores.
- Contamination by iron bacteria – these micro-organisms, which occur naturally in moist sediment, may already exist in the aquifer or may be transported to the bore during the drilling process. They create a slime which can clog equipment, corrode the bore casing, and discolour and contaminate water. Disinfection of all new bores is recommended, using liquid chlorine or a proprietory chemical produced specifically for bore disinfection.
- Contamination by pollutants, either from surface water entering the aquifer or below-ground contamination. Pollutants include septic wastes, fertilisers, pesticides and other chemicals, wastes from intensive animal industries, land fills and stockpiles, abandoned bores and mines. When siting a new bore, consider the proximity of possible sources of pollution, including previous land uses, and avoid placing the bore at the bottom of a gully where surface runoff can submerge it.
- Blockages in the bore casing or screen, which prevent water entering the bore. Blockages can be caused by corrosion, fine sediments and bacterial slime.
- Pump malfunction – for new bores, ensure the capacity of the pump is not greater than the yield of the bore.

Farm dams

A well-constructed farm dam will provide adequate water in most seasons at an economical cost.

Planning a dam

Work out the dam size

- Estimate your water requirements. These will depend on the geographic location, crop type, type of stock and stock numbers (see Table 7). Include an estimate of evaporation losses – up to 30% on the coast and 50% for inland areas.
- Estimate the storage requirements – how long the water will have to last without replenishment – 12 months duration may be sufficient on the coast and two to three years in dry areas.

Choose a dam site

- Look at the farm's topography – on undulating land, a gully is a good site because it requires minimal earthworks, and hence costs less. On gently sloping land, a hillside dam is suitable; on flat land, an excavated tank can be constructed.
- Consider the catchment yield – the catchment is the area that collects rainfall runoff and channels it into the dam. The ideal catchment area has sparse vegetation and a

hard surface (eg roads, rooftops or stony soil) that allows the runoff to flow over the surface into the dam. Deep soils covered with lush vegetation quickly absorb rainfall and often yield minimal runoff. If the catchment does not provide sufficent runoff, catch drains can be constructed. These drains collect runoff from outside the catchment area and direct it to the dam.

- The capacity of a small gully storage dam can be estimated by the formula:

$$\text{Volume} = \frac{\text{width x maximum depth x length}}{5}$$

- The capacity of a hillside dam can be estimated by the formula:

$$\text{Volume} = \frac{\text{surface area x maximum depth}}{3}$$

Source: *Planning Your Farm Dam*, Rural Water Advisory Services, Queensland Department of Natural Resources, July 1995

Check licensing requirements

In many cases you can build an earth dam without restrictions but always check with your local council and water authority before proceeding with the construction.

Test the soil at the site

The embankment must be structurally stable and able to hold water. A soil test will determine whether the natural soil is suitable – the ideal soil is a clay which is impermeable and stable. (NB: clays vary in their characteristics, not all are suitable for dam construction.)

To test the soil, obtain samples by drilling auger holes or digging test pits with a backhoe. Obtain samples from the embankment centre line, the bywash and the gully bed.

Livestock water requirements

The following figures are only yearly average estimates. Requirements can vary according to climatic conditions, the amount of work the animal is doing, and the variety of animal concerned.

Table 7 Water needs for livestock

Type of Livestock	Estimated annual (KL per head)	Daily (L/head/day)
Ewes on dry feed	3.6	9–10
Mature sheep – dry feed	2.7	7.0
Mature sheep – irrigated	1.35	3.5–4
Fattening lambs – dry feed	1.2	3.3
Fattening lambs – irrigated	0.6	1.7
Dairy cows in milk	33	90
Dairy cows – dry	20	55
Beef cattle	17	45
Calves	8.2	22

Table 7 Water needs for livestock (continued)

Type of Livestock	Estimated annual (KL per head)	Daily (L/head/day)
Horses – working	18	50
Horses – grazing	13.5	37
Pigs – brood sows	8.2	22–30
Pigs – mature	4.1	11–15
Poultry – laying hens	12 per 100 birds	25–32 per 100 adults
Poultry – pullets	6.3 per 100 birds	17 per 100 adults
Turkeys	20 per 100 birds	55–60 per 100 adults

Other requirements

Wash Down Requirements	
Piggeries and dairies	50 000 litres per 10 sq m
Domestic Requirements	
For family of two	200–270 litres/day
For family of four	270–340 litres/day

Maximum salinity for farm livestock

Poultry and pigs	2000 ppm
Dairy cattle	3000 ppm
Sheep, beef, cattle or horses	4000 ppm

(Animals may tolerate double these levels for temporary periods during drought.)

Sources: *Farm Management* by John Mason, published by Kangaroo Press; Landcare Note SC/007 from the Victorian Department of Conservation and Natural Resources.

Problems with water

Mosquitoes

Mosquitoes and other undesirable insects can breed in still water or moist places around a farm. In areas where serious mosquito-carried diseases (eg malaria, Ross River virus) are common it is extremely important to keep these insects in check. Fish or other insect-eating animals in the water will help reduce their numbers. If the water is chemically treated or sprayed periodically this can also keep insects at bay.

Willows and waterways

Willows (*Salix* species) are commonly found growing along waterways in many parts of the world, including temperate Australia. While these plants are excellent for preventing erosion of the banks of dams and rivers, they can cause significant and undesirable changes to the ecology of the watercourse. Willows, unlike most other vegetation, can spread their roots into the bed of a watercourse, slowing the flow of water and reducing aeration. Willow leaves decompose much faster than many other types of leaves, creating a flush of organic matter in autumn when they drop and an under-supply for the remainder of the year.

Research at the University of Tasmania has shown willows have a negative effect on populations of invertebrate animals.

Algal blooms

Algae are small forms of plant life that thrive in moist, light and fertile conditions. Still, sunlit water, such as that found in dams, lakes, troughs and open storage tanks, stimulates the growth of algae. Runoff from fertilisers, especially those containing nitrogen and phosphorus, further encourages growth, to the point where the water becomes unpalatable and potentially poisonous to livestock, humans, fish and other aquatic organisms.

Several species of blue-green algae are toxic. A bloom of blue-green algae will discolour the water, turning it an acidic green colour. It may have an unpleasant odour. The bloom can develop very quickly – in less than a week – making the water unsuitable for irrigation and for watering livestock. As the bloom decomposes, it reduces oxgen in the water, and fish may die. Even after several months, the sun-dried scums can remain toxic to animals.

The best way to control algal blooms is to prevent them happening. Minimise nutrient runoff into dams by avoiding excessive fertiliser use on the farm; fencing out stock from dams (use gravity-fed troughs for drinking water instead); establishing buffer strips of vegetation (grasses, trees and shrubs) to help stop nutrients and eroded soil entering the dam; and avoiding the domestic use of washing powders and detergents containing phosphates.

Artificial aeration helps to control blooms by mixing water layers and increasing oxygen levels. The simplest method is to cascade the water into a holding tank or dam.

Algal blooms can be treated in dams (but not streams or natural waterways) with algicides but they must be used with caution – the algicide must not affect groundwater or catchment areas. Consult a farm advisory officer for advice.

Livestock contamination

Canadian research has shown that farm productivity can increase if grazing animals are fenced away from watercourses running through a property. Stock should not have direct access to creeks or rivers. The research showed that the quality of livestock drinking water has a direct bearing on livestock health and profitability. Hence, don't allow water to be fouled, and the farm will be more productive! Significant reductions have also been noted in streambank erosion as a result of decreased trampling by stock.

Source: Land and Water Resources Research and Development Corporation, Research by Dr Walter Williams *et al.* seen in *Acres Australia* Vol 3 No. 6.

Flood

Excess rainwater runoff can be a cause of severe difficulty to the farmer, resulting in erosion and loss of valuable topsoil. Floods can also cause severe losses through death, or reduction in health of stock, damage to fencing and structures (eg sheds, bridges), temporary reduction in area for stock to graze, and boggy conditions for movement of stock and machinery.

There are some simple means by which flood damage can be minimised. These include:

1 Ensuring that any structures such as sheds and shelters, and stored food (ie hay and silage) are located as high as possible above natural flood plains.
2 Soil that has vegetative cover will always stand up to flood better than bare ground. Overgrazing or cultivating soil at times of the year when floods are likely increases the potential for soil loss if flooding occurs.

3 If possible, arrange fencing of low-lying land to include a few areas where stock can retreat as water rises.
4 Have a procedure for evacuating stock in case of flood, including:
   - Having suitable transport available (boats may be necessary in regularly flooded areas)
   - Having a suitable place to take stock, which has temporary provision for food, shelter and water
5 Regular monitoring of flood levels - don't leave it too late to act.

Water quality

Water quality is affected by the type and amount of impurities. Physical impurities are particles in the water; chemical impurities are substances dissolved in the water. Biological impurities are living organisms such as algae and some micro organisms. Bacteriological impurities are shown separately because of their importance to human and animal health.

Rain or creek water is unlikely to have serious physical or chemical impurities, but may develop algal problems, particularly if exposed to light and if nutrient levels are high. Bacterial impurities may develop if this water is stored improperly.

River or spring water is unlikely to have biological impurities (eg algae), but may have chemical, physical or bacteriological impurities, depending on the source.

Bore or channel water hardly ever has physical or algal impurities, but may contain salts (causing hardness). Bore water may also contain iron.

Dam and irrigation water generally contains few chemical or biological impurities if properly managed, but may have sediment or other physical impurities and may develop medium levels of bacteria, particularly if animals are allowed to foul the water.

The quality of water may be found by testing a sample. This is normally carried out by such organisations as:

- Companies that sell equipment for the treatment of water
- Local organisations such as dairy factories and water treatment trusts
- Departments of agriculture, primary industries or similar bodies
- Departments of mines or similar bodies
- Departments of health
- Water supply authorities

Before collecting water for testing you should contact the testing organisation you have selected for advice on how the sample should be collected.

Salinity

A major concern with water quality is its level of salinity. Salinity in irrigation areas in many dryland countries, including large tracts of inland Australia, has been the cause of severe environmental and economic degradation.

As salinity levels rise in an area, the productivity potential falls. Salt-affected soils suffer from surface crusting, reduced infiltration and restricted subsoil drainage. Crops and pastures exposed to saline irrigation water experience water stress, resulting in leaf scorching, leaf fall, slow growth and reduced yields. In extreme cases, vegetation dieback occurs and the soil is left exposed to erosion.

Testing water salinity

The level of salinity in water can be measured by testing for electrical conductivity (EC). Small hand-held EC meters are readily available at reasonable cost. Regular tests should be conducted on the farm water supplies to determine their suitability for livestock and irrigation.

Treating saline water

Short-term options

In the short term, little can be done about excessive salt in a water supply without significant cost. Some of the options are:

- Mixing saline water with non-saline water, if available
- Applying extra non-saline water to the soil to leach salts below the root zone; good subsoil drainage is required to ensure the leached saline water is removed from the topsoil
- Desalinating the water using a treatment plant, small plants are available but they are expensive to purchase, and have high operation and maintenance costs. Some problems that may occur with desalination treatments include the need for water pre-treatment (using sand filtration, micro-filtration or UV treatment), the difficulty of treating water with high iron, silica or manganese, and the problem of disposing of the residual saline concentrate.

Management options

- Choose salt-tolerant plant varieties (see below)
- Use mulches under crops to reduce surface evaporation, which results in a buildup of soil salinity
- Change fertilisers – fertilisers contain varying amounts of salts (described as a 'salt index'), and it may be possible to use a fertiliser with similar nutrients but with a lower salt index; eg potassium chloride has a salt index of 114, while potassium sulphate has a salt index of 46
- Use drip irrigation in preferance to other forms of irrigation – the benefits of drip irrigation include minimal evaporation and a reduction of the effects of salinity by maintaining a continually moist soil around the plant roots and providing steady leaching of salt to the edge of the wetted area

Long-term strategies

Over time, large-scale planting with tolerant tree and pasture species can reduce salt levels in the soil by lowering the water table. Tolerant trees can be planted directly on salt-affected land or above shallow saline groundwater in the recharge area.

Salt-tolerant trees

Acacia dealbata, A. mearnsii, A. melanoxylon, A. stenophylla, Allocasuarina cristata, A. glauca, Casuarina cunninghamiana, C. obesa, Corymbia citriodora subsp. *variegata,*

Corymbia tesselaris, Eucualyptus camaldulensis, E. occidentalis, E. sargentii, E. spathulata, Melauleuca halmaturorum, M. leucandendra, M. uncinatum, Pinus pinaster, P. radiata, Phoenix canariensis, Tamarix spp.

***Salt tolerant shrubs, grasses and pasture* spp.:**
Atriplex spp., *Elytrigia elongata, Halosarcia* spp. *Melaleuca nodosa, Paspalum vaginatum, Puccinellia ciliata, Trifolium michelianum*

A number of techniques are used to establish plants on salt-affected soils:

- Good drainage is desirable to prevent continued buildup of salt, ideally trees and shrubs should be planted on mounds up to 50 cm high
- Build the mounds several months before planting to allow some salts to be leached prior to planting
- Apply heavy mulches around each plant to reduce evaporation that leads to salt accumulation on the soil surface
- At planting time, scrape away the top 2 cm of soil and plant trees and shrubs into deep holes on the top of the mound
- Grow salt-tolerant grasses and legumes over the site to increase water uptake but make sure they don't impede the growth of the trees and shrubs
- Don't let animals graze the planting site

For more information on salinity see Chapter 3.

Tastes and odours

Many tastes and odours that commonly spoil the palatability of water are caused by mineral or organic substances dissolved in it. They indicate pollution of the water supply. The pollution may come from algae, fungi, bacteria, animal waste, decaying organic materials, metallic compounds such as iron and manganese, chlorides, hydrogen sulphide, sulphates, industrial waste and sewage.

Treatments and remedies:

- Locate and remove the source of the taste and odour
- If caused by algae, treat as described later in this chapter
- Depending on what is causing the problem, chlorination may be necessary to kill bacteria and make the water safe to use
- Commercial water treatment companies market activated carbonless filters which will remove taste and odours
- Aeration treatment may remove tastes and odours caused by iron

Reed-beds

The purification of waste water and effluent using reed-beds has been successfully achieved for hundreds of years. By allowing dirty water to pass through wetlands planted with reeds and rushes, the roots of certain plants release oxygen, which helps micro organisms break down and filter out impurities. The method can ultimately produce high quality water which may be suitable for drinking. The plant biomass that grows in this system can also be harvested as a source of mulch, or perhaps as a crop in its own right.

Reed-beds may be naturally formed wetlands or artificially constructed and planted channels and beds. Given the current degree of environmental pressure on the few natural wetlands remaining, it would appear that further pressure on or usage of such wetlands is unwise. However, the deliberate building of new, well-designed wetlands/reed-beds could be a very useful enterprise, especially for treatment of effluent from dairy farms.

When micro organisms break down water pollutants, they use up oxygen. This oxygen consumption varies with different materials, and is known as the biological oxygen demand (BOD). For example, nutrient-rich wastes such as farm manures or silage effluent have a high BOD. When these pollutants find their way into waterways, the oxygen level in the water becomes seriously depleted as a result of breakdown processes, causing parts of the natural flora and fauna of the waterway to die. When the water body is small and the flow rate is slow (eg in conditions of low rainfall), this problem can be quite severe. The blue-green species of algae are then able to flourish, poisoning and fouling the water even further.

The problem of limited oxygen supply may be overcome by the use of structures such as pebble streams, rock-lined channels or waterfalls. In this environment of plentiful oxygen, micro-organisms such as bacteria, yeasts and fungi become established and thrive on the surfaces of the pebbles or rocks and consume the soluble polluting matter.

Alternatively, plants may be used to supply the oxygen necessary for micro organisms to break down pollutants. Some plants, mainly reeds and rushes, absorb atmospheric oxygen through their leaves and transfer it down hollow stems to their extensive root systems. The oxygen is then released through fine root hairs into the soil where it helps build up micro organism populations and facilitates the breakdown of organic matter. Reed-beds work most effectively when a dense layer of rhizomes and root hairs is formed. This may take about three years to fully develop.

Water saving measures

There are ever-increasing demands for what is essentially a limited resource – water. This increased demand leads to the construction of more water storage facilities which have a heavy impact on the environment, in such ways as flooding valuable agricultural land or native forests, or by changing the natural pattern of water flow in streams which have been dammed. By minimising the amount of water we use, we can reduce the requirement for additional water storage facilities and therefore reduce the likelihood of negative impacts on the environment, as well as possibly reducing our production costs.

Most of the following methods of conserving water can be applied equally to crop production or to home garden use:

- By choosing plant species and varieties that best suit the local climate
- By maintaining a well balanced fertile soil appropriate to the plants selected
- By watering in the cool of the day
- By using micro-irrigation systems, eg trickle systems, where possible – these are much more efficient in their use of water than other irrigation systems
- By slow, thorough watering – a thorough deep watering once or twice a week will be more effective than light waterings every day or two

- By avoiding spraying water on windy days.
- By considering soil type when selecting a watering system – for instance, clay soils hold water well and will distribute it horizontally, so a drip system is suitable, whereas water runs quickly through sandy soil, so a micro-spray would be more suitable as it distributes water over a broader area.
- By reducing excess evaporation – this can be achieved by keeping bare soil covered, using mulches or plants; both organic (eg bark, compost, lucerne) and inorganic (eg gravel) mulches are excellent for reducing evaporation; compact groundcovers will slow evaporation from the soil but they will use a lot of water themselves; larger plants will shade the soil and limit evaporation but they can make getting water to the soil in the first place rather tricky
- By using rainwater tanks to gain extra water, particularly for domestic use, and for collecting water from large sheds to water stock – this can reduce the need for installing water mains to some areas to provide water for stock, troughs can be filled directly from the tank

Recycling household water

It is possible to use excess water from the house to water gardens, in particular water from showers, baths and washing machines. This can reduce the demand on water needed for other uses (eg watering stock, irrigating crops).

Re-using water from the house will involve some plumbing to reduce the drudgery of bucketing water out onto the garden. The simplest method is to undo your drain pipes and let the water from sinks flow into a bucket for smaller amounts, or connect a hose to the drainpipes and let the water flow into a holding tank. This water is referred to as 'greywater' and can contain soaps, food scraps, grease and bacteria.

Water with cleaning liquids and solvents that are harsh to the skin or harmful to plants should be diluted before being used in the garden. Do not use water from the dishwasher. You should be careful to use biodegradable soaps and avoid detergents with boron. When added to the soil such detergents may be toxic to plants.

Use trickle irrigation to apply greywater, as wetting the leaves with it may cause leaf burn. A filter will be necessary to make sure any solid materials or residues in the greywater do not block the pipes and nozzles. Another method is simply to allow the water to run across the ground surface (flood irrigation) by pouring water out of a bucket or allowing it to run out of a hose. Remember to water different areas each time to get even coverage.

You should check with your local council to confirm that they allow the use of greywater.

Using farm/waste water

Treated/purified water from activities such as dairy washings can be utilised to water crops or pasture. This treated waste can often have high levels of some valuable nutrients (eg nitrogen, phosphorus), reducing the need for fertiliser applications. The use of these nutrients also reduces the likelihood of them entering streams and causing nutrient loading. Ideally, before using such treated waste water it should be tested to ensure that no elements are present that might cause toxicity problems (eg heavy metals).

Sustainable agricultural aims to maintain soil health and prevent large scale degradation such as erosion.

An all-round soil conditioner and organic fertiliser containing seaweed extract.

Soil profile – note the organic matter content at the top.

A soil treatment and clay breaker designed to improve water penetration and reduce surface runoff.

A Permaculture system in Queensland, Australia.

Worms can be used to improve soil condition or to process compost prior to use.

Finished compost is an excellent soil additive. Whilst it incorporates organic matter into the soil, it will not draw nitrogen out of the soil in the same way that raw organic matter will.

Mouldboard plough – can be useful for incorporating organic matter into soil that has been cultivated for many years.

Stable manure stockpiled for composting.

This small modern compost bin is suitable for composting small pieces of organic matter and household waste.

Pasture seed drill.

Dung beetle at various growth stages. In addition to dung removal, dung beetles enrich the soil and reduce numbers of dung-feeding flies.

Vegetables being grown in conventional monoculture.

Flood irrigation. The mown grass cover between tree rows helps to prevent erosion.

Hydroponic lettuce. Hydroponic growing allows excellent control over both production and farm wastes

Irrigation channel.

Flood irrigation outlet.

Windmills provide a clean and cheap way of pumping water on the farm.

Swales (for water catchment) in Queensland, Australia.

Water tanks should lock out light to discourage algal growth.

Farm dam.

Trickle irrigation used for establishing fruit trees.

Many tastes and odours that commonly spoil the palatability of water are caused by mineral or organic substances.

Soil electro conductivity (EC) meter. The EC reading indicates how fast electrons are flowing in the soil. The faster the flow, the higher the nutrient availability of the soil.

Treed paddocks provide shelter for stock and protect against erosion on steep slopes.

A natural pest trap. An attractant placed within the container will lure pests to the poison.

Poorly drained pastureland. Areas like this are a breeding ground for pests and diseases.

The plastic strings above this garden move in the wind and repel birds.

Predatory mites eat other pest mites. They are bred commercially and farmers can release them onto crops as a biological control agent.

Weed control is essential. In this experiment, part of a row crop is swamped by weeds when left with no weed control effort.

Basil – for use as a companion plant.

Tansy is used as a natural fly repellent in companion planting.

Wasting water

Water wastage during irrigation has been a major problem for irrigators in the past. There has been a good deal of research in order to rectify the situation. Wastage occurs in numerous ways, including evaporation, seepage and runoff. The key requirements of water use are efficiency and even distribution, so that it is being utilised by the plants in order to aid quick, healthy, even growth. Problems such as waterlogging or salinity should be monitored and in many cases can be avoided with good irrigation management techniques.

Types of water wastage

Evaporation

A certain amount of water loss through evaporation is inevitable. Water that is stored in ponds and lakes is more susceptible to evaporation due to large open surface areas. Flood irrigation, too, will have more severe evaporation losses than trickle or drip irrigation. In all cases, irrigation that is undertaken at night will suffer less from evaporation losses. Evaporation also takes place through the plants that are being irrigated. This is referred to as transpiration and as a natural process of plants cannot strictly be viewed as water wastage; however it is an important factor in estimating crop irrigation requirements.

- Evaporation – the loss of water as vapour from a free water surface
- Transpiration – the loss of water as vapour, generally through the stomata of leaves
- Evapotranspiration – the combination of the above two factors which is essential when estimating irrigation levels for crops

The loss of water through evaporation is commonly controlled, either by the type of irrigation employed, or by the timing of irrigation practices in relation to local climatic conditions.

Seepage

Seepage is another factor which contributes to water loss. It, too, is impossible to check completely. Seepage occurs through the base and walls of canals and dams, which are usually constructed from locally available soils. The degree of compaction and permeability of these soils is what accounts for the levels of seepage. Channels or dams can be lined to reduce seepage, particularly for soils with high permeability levels. Materials such as bentonite can be added to dam water. This material is a type of clay that settles to the base of the dam and swells, helping to seal the dam and reduce seepage.

Runoff

Runoff is the result of water reaching the soil surface faster than it can infiltrate into the soil. This may be as a result of poor irrigation practices or due to excessive rainfall. It may be further intensified by poor drainage. Irrigation is the most controllable factor of water wastage, but all too often is not given the consideration it deserves. Many variables determine the optimum irrigation rate. These include soil type and quality, climate, soil suction levels (which can be tested using a tensiometer), particular crop requirements, and the recent watering/rainfall history of the area to be irrigated.

Runoff can be re-used if the appropriate drainage and recycling techniques have been included in the irrigation design, thus minimising wastage. Likewise, runoff from excess rainfall can also be used, if the water can be diverted to a suitable storage (eg dam), or if storage areas are placed where natural flow of runoff can be collected (damming a natural drainage basin). Careful placement of dams can ensure that as much of the runoff from your property as possible is collected for later use. This can also help reduce impacts on the environment as nutrient or chemical laden water from eg animal wastes and fertiliser and pesticide applications is collected before it reaches streams or lakes.

Runoff water can often be used on the property without treatment, but if pollutant levels are high it may require some treatment (eg through a reed-bed system).

Overspray

This mainly concerns the use of overhead sprinklers, but misters could also be included. This is where the water has been sprayed in areas where it is not needed. Careful placement of sprinklers and adjusting the arc of spray of each sprinkler, as well as the amount of water being sprayed from each one, can significantly reduce wastage.

Scheduling

Often the water used in irrigation is scheduled. For instance the farmer may have talked to the water bailiff and ordered so many megalitres to be available for a certain day of the week. The water bailiff then releases that amount of water into the river or canal system. There are times, however, when the water is not required due to a heavy local rainfall, or evaporation rates are markedly less than expected. In this instance the water that is earmarked for the irrigation is not going to aid the crop, and might possibly even hinder its growth. It will still have to be paid for and will have been wasted.

The bailiff may be able to allocate the water elsewhere, but the main strategy to avoid this scenario is to avoid scheduling, until you as the irrigator are certain of when and how much water you require. This is very much a good management issue and can be especially critical in times of uncertain water allocation due to drought, and/or increased demand.

Recycling waste water

Waste water that has been used for irrigation, domestic and commercial use can be treated and recycled and made fit for use for irrigation purposes. The water is treated in a number of ways, including filtering or use of settling ponds for removal of solids, biological treatment and disinfection. Misconceptions still exist, however, over the actual quality and health aspects of treated water and, as a consequence, much of this treated water – which is often high in nutrients – is flushed into our waterways, where it contributes to environmental problems such as algal blooms. This same high nutrient content water brings positive benefits from an irrigator's viewpoint as it contains valuable nitrogen and phosphorus.

Water recycling can also be practised on a more localised scale. Water that has already been used for irrigation or other purposes (hosing down cattle yards) will have deteriorated in quality due to the absorption of solids such as fertiliser, organic matter, etc. This water is often best if allowed to settle in a dam, allowing solids to sink to the bottom, at which time the water will again be of a useable quality.

The most common approach to recycling excess water is to design channels and canals in such a manner that drainage channels flow into primary channels or dams that service other areas to be irrigated. This system, to be at its most effective, needs to have good design of its levels and gradients.

Swales and keylines

We can make better use of limited water resources by first understanding the way rainwater runs across a slope, and then reshaping the land to control the water flow – both where it goes and the speed of flow. Yes, speed does matter. When water flows fast over a hard soil, it doesn't soak in to any great degree, and erosion may be high; but when it flows more slowly, more water will soak (infiltrate) into the soil, and the potential for erosion will be reduced.

These considerations are the foundation for sustainable farming practices which have different names in different places.

Swales

This is an old concept, practised in many parts of the world and promoted strongly by permaculture practitioners. It involves the creation of long, level hollows, furrows or other excavations, or barriers created across a slope (such as a long pile of rocks or rubble). They are used to intercept overland water flow and then hold it for long enough to let it slowly infiltrate into the soil.

It is important when creating these swales that their bases are treated in such a manner (eg ripping, cultivating, adding soil ameliorants) that helps improve the infiltration rate of water into the soil beneath.

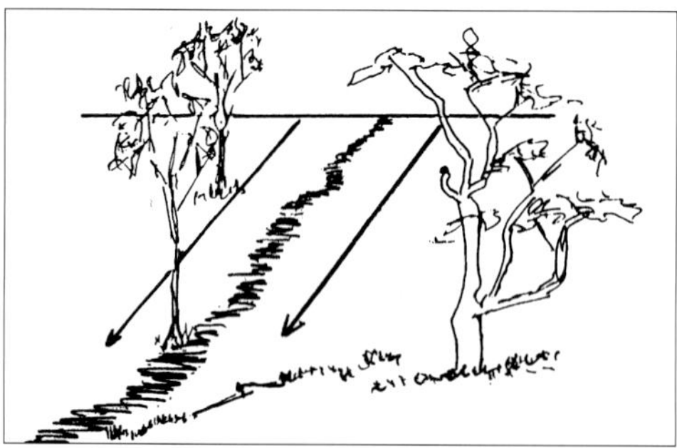

Figure 4.1 Water flows down a slope before contouring or swales.

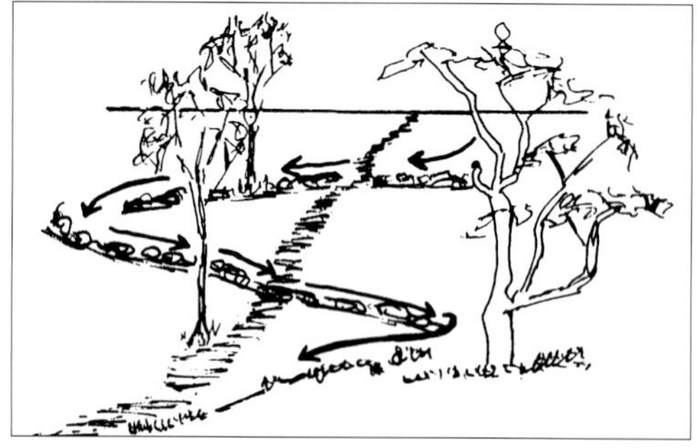

Figure 4.2 Redirected water flow is slower and more beneficial after swales or contouring.

Figure 4.3 Swales will intercept runoff water to allow for improved filtration and reduced erosion. Growth is usually faster on the ridge of swales, due to available water and dissolved nutrients.

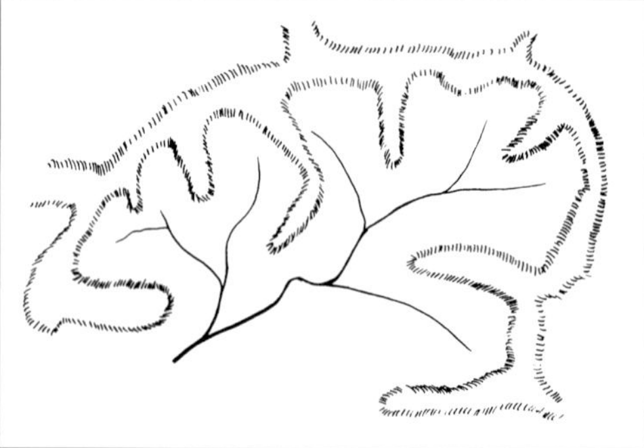

Figure 4.4 The shape of the ridge and valleys give rise to the watercourse. Water tributaries are made up of smaller valleys which run off from the ridges and combine to form the watercourse.

Swales can be stabilised by planting them out with pasture species and/or planting their slopes with trees and shrubs. Infiltration rates will generally increase with time due to the effects of tree roots on the soil in the swale, and through humus accumulation. Swales are also effective in trapping eroded sediment.

Keyline design

This is a concept developed in Australia by the Yeomans family, incorporating the natural contours of the land in order to plan the positioning of dams, treelines and irrigation layout.

Designing a keyline system for any property requires good initial planning. A contour map is essential in order to best understand the rise and fall, and flow of the land. The leading proponents of keyline design recommend the use of orthophoto maps (aerial photomaps which are available from keyline consultants) because of their increased contour detail.

Planning should, if possible, be undertaken before the land is acquired as this will enable the viability to be assessed before substantial investment has been made. All land can be improved through the use of keyline design although obviously some parcels of land will be better suited for productivity purposes.

Three fundamental concepts that must be understood are keypoints, keylines and keyline cultivation patterns.

1 Keypoint refers to a position located along the centreline at the base of the steepest part of a primary valley.
2 Keyline is a line that runs through the keypoint and extends to where the contours of the valley start to become the sides of the ridge.
3 Keyline cultivation patterns. The basic rules of thumb are that cultivation on ridges (above the keylines) should be parallel to and above the contour lines, whereas cultivation below the keylines should also be parallel, but below the contour lines. This means that water runoff above the keylines is directed towards the more gradual slopes for slower dispersal into the soil, and below the keylines it will be directed towards the greater slope for quicker dispersal, and hence will not result in swampy, and therefore possibly saline, conditions.

The use of keyline cultivation is extremely beneficial for effective flood irrigation practices. It allows for inexpensive flood irrigation of undulating land as well as fast flood irrigation of flat areas.

This is a crucial point because generally with flood irrigation the water tends to lie for greater periods in flats while it is being coaxed to cover the entire area. This means that important aerobic microbial organisms are deprived of oxygen in these lower areas, also contributing to the poor quality of the land in low areas.

Keyline design is initially concerned with the topography and climate of the land parcel, but also incorporates the use of treeline,s and plans the positioning of dams, irrigation channels, fences, farm roads and buildings. Planning should take place in the following order as proposed by Yeomans in articles on keyline design:

1 Water
2 Roads

3 Trees
4 Buildings
5 Fences

Water includes calculation of key points and keylines in order to aid placement of dams and irrigation channels. Placement and size are decided by the contours and the surrounding suitability of flood irrigation land.

Roads should not cut across contours as this interrupts water runoff flow. Generally, road placement should be along the top of ridge lines and elevated contours.

The importance of treelines on a property cannot be understated. It is an area of farm management that has become increasingly scrutinised in order to achieve the best results without loss of production or headaches during cropping or related farming methods. Trees are crucial to sloping land in stabilising the soil, providing shelter from the elements and controlling erosion. It is these same principles that make them an integral part of keyline design.

Buildings are placed by considerations of comfort, aesthetic attributes and practicality.

Fencing can be applied to correlate with the planned subdivision of the property based on its production requirements. Fencing should also be used to protect trees and dam access from damage by stock.

Irrigation systems

Irrigation system design

Before designing an irrigation system you must first consider:

1 Water availability
   a What does it cost?
   b What quantities are available?
   c Is it available at all times of the year?
   d Do you need to build dams or holding tanks to cope with the times of restricted water supply?
2 Source and quality of water
   a Is it clean?
   b Does it carry salt, sediment, pollutants, diseases, etc?
   c What pressure is it supplied at, or do you need to pump it (if so, how far?)
   d Is it from a canal, river, creek, bore, municipal supply, dam?
3 Regulations
   a Are there any restrictions on use of water?
   b Are there any restrictions on building dams, canals, etc?
   c Are there any drainage regulations?
   d Check with the local council.
4 Site details
   a You need to know the shape of the site, levels, drainage patterns, locations of buildings, fences, roads walls, etc.
   b You need to know the location, type and quantity of plants to be watered.

5 What money is available to create the system?
6 What manpower and level of expertise are available to operate the system?

The importance of system design cannot be overly stressed. The effectiveness and overall efficiency of the irrigation project will be largely dependent on system design. Often design is approached from a narrow focus that addresses only certain criteria. While achieving much of the required aims, such a system lacks the ability to diversify, to cater for future expansion, or ignores minor design issues which have the potential to develop into larger problems at a later date.

Design processes can seem initially costly, as good design often requires time – time to evaluate the individual complexities of the project; time to put down potential options by way of rough plans and scenarios, and, finally, more time to review these options and hopefully select the optimum design for that particular irrigation system. Even small-scale irrigation systems are generally costly and intended for long-term usage, therefore it is important to get the design process right!

Steps in the design process:

1 Initial aims and objectives
2 Research and consultation
3 Primary options and design plans
4 Scenario testing/comment
5 Secondary options and plans
6 Design of an appropriate irrigation system

Issues such as the actual type of irrigation method to be selected will be looked at during the design process. In most cases there will be more than the one option and the final decision will be decided on economic and technical feasibility.

Maintenance procedures and scheduling

Irrigation efficiency is sustained through proper maintenance and use of a system. Maintenance may involve tasks such as travelling around the farm on a motorbike, shovel at hand to deal with any breaches in channel walls, or an intensive program of oiling and testing every aspect of the irrigation process, usually during the off season.

Further to this, sprinklers wear, pipes corrode or in the case of plastics can develop punctures or breaks at the seals or joins. Water pressures can be relatively easily checked, whereas pumps might require specialised attention.

Maintenance of equipment is carried out to prevent costly breakdowns. Maintenance programs can be divided into three groups:

1 Periodic equipment inspections to find out conditions that could lead to breakdowns or too much depreciation
2 Upkeep to remedy such conditions
3 Contingency work

These activities may include proper lubrication, job planning, and scheduling of repairs. This can lead to fewer breakdowns which means:

- a decrease in production downtime

- fewer large-scale or repetitive repairs
- lower costs for simple repairs made before breakdowns (because less manpower, fewer skills, and fewer parts are needed)
- less overtime pay on ordinary adjustments and repairs than for breakdown repairs
- less stand-by equipment

Periodic inspections

Periodic inspections are the best way to achieve a preventative maintenance program. Thus, unsatisfactory conditions are discovered and corrected in their early stages. Frequency of inspections depends upon the amount and degree of use and will vary from one area to another. The best frequency cycle can be established by analysing the equipment based on:

- age
- condition
- value
- amount of use
- safety requirements
- number of hours or mileage operated
- susceptibility to wear, damage, and loss of adjustment

These inspections should be carried out at a garage or workshop by personnel trained in this kind of work. During these inspections, every moving part of the equipment should be checked and worn parts removed and replaced. Following inspection, the equipment should be thoroughly cleaned, lubricated and tested.

Routine upkeep

Routine upkeep of equipment includes such chores as lubrication, oil changes, battery checks and replacing seals or washers. Most of these chores should be the responsibility of the operator, who should report any problems to the person in charge or property manager as soon as possible.

Contingent work

Contingent work includes the repair of major breakdowns and general overhauls of equipment. At the same time a regular inspection should be completed and all worn and broken parts removed and replaced.

Scheduling work

Maintenance work can be scheduled by using overall charts for all equipment and individual cards for each piece of equipment. The overall chart gives a quick picture of the workload. This chart lists the days or months across the top and an itemised list of equipment down the left side. A system of symbols can be devised to show different types of repairs and adjustments to be made to each piece of equipment. The relevant symbols are marked under the date the activity is to be carried out, and opposite the piece of equipment that is to be worked on.

The individual cards for each piece of equipment allow for greater details to be recorded. All repairs, both major and minor, and maintenance should be recorded on the card, and should include a list of materials/parts used and time required.

Surface/flood irrigation

This is the most widespread form of irrigation and, if land and water conditions are favourable, also the most cost-effective irrigating option. Land is graded at an appropriate slope and gravity is used to distribute the water across the surface of the land to be irrigated.

Border check system

With this type of irrigation a field is divided into bays, all running downhill away from the source of water. These bays are divided from one another by a raised ridge of earth known as the check bank. Water is let onto the paddock through a bay outlet, allowed to flood about two-thirds of the way down the bay and then closed off; the desired result being that the lower third will receive the surface drainage from the higher part of the bay and runoff will be minimised.

This type of irrigation is not suitable for soils that have high absorption rates, such as deep sandy loams, as water use will be too excessive. Border check irrigation is used extensively for pastures, lucerne and fodder crops. It is very common in some dairying districts upon which soldier settlements were created after the two major wars last century.

Relatively new developments in flood irrigation techniques have greatly improved the efficiency of this type of irrigation. Better land forming and more accurate grading using laser levelling equipment, new design concepts, recycling of drainage water and better dethridge wheels all contribute to more efficient flood irrigation. (A dethridge wheel is a device which moves and measures water taken from irrigation channels.)

This system may be used in orchards or vineyards, but is not as commonly used in such situations as furrow irrigation. It involves creating a series of side-by-side bays, usually rectangular, which are fed water from a channel running along the high side of the line of bays. Progressively, one at a time, an opening is made in the channel wall at the top of each bay, allowing that bay to flood. Once flooded, the opening is closed, and the next bay can have an opening made. The bays and channels are formed by mounded earth walls. The gradient across the bays must be carefully engineered to allow proper distribution of water.

Hillside flooding

Also called 'contour ditch irrigation', this system is used in hilly country which is too steep to allow other forms of flood irrigation. This system can be used on land with slopes up to 1 in 10. A supply water channel runs along the top of a slope. This channel is opened progressively – one point at a time – along its length, allowing water to spill down the slope. The slope may need earthworks to ensure an even surface gradient, and hence an even distribution, of the flooding water.

Furrow irrigation

This is the most common flood system for orchards and vineyards. Water is released from a supply channel on the high side of the area to be irrigated into furrows or channels. If vines or fruit trees are planted on mounds, a natural depression is made between the rows. This can be ploughed to form the required furrows, or the depression as it exists may be used as a furrow.

Sprinkler irrigation

The following characteristics of sprinklers can affect the efficiency of irrigation systems:

Wind velocity and wetting pattern

The basic wetting pattern of a single-nozzle sprinkler is roughly conical in shape. Winds over 8 km/hour in velocity will distort this pattern, giving an uneven ellipse patterned distribution of water. High-pressure sprinklers with long trajectories are the worst affected by wind velocity.

Allowance for wind distortion can be made by decreasing the spacing of sprinklers perpendicular to the wind direction, providing of course that the direction of the wind is reasonably constant. The following table indicates the effect that wind has on water distribution:

Table 8 Effect of wind on water distribution

Wind velocity (km/hour)	Sprinkler spacing perpendicular to wind direction as a % of wetted diameter
0	65
0–6	55
7–13	40
Over 13	30

Drop size

Drops greater than 4 mm in diameter have a tendency to damage delicate plants and contribute to water erosion problems, whilst drops less than 1 mm diameter are easily deflected by wind. Medium and low pressure sprinklers mainly produce drops within the 1–4 mm diameter size range while rain guns tend to produce a wide range of drop sizes, with a large proportion at or above 4 mm.

Rotational speed

Sprinklers which have very low rotational speeds tend to cause a breakdown of the surface structure of fine soils, due to the action of long periods of wetting followed by long periods of dryness. At high rotational speeds, distribution becomes uneven. A speed of 2.5 m/s at the perimeter of the throw will give the most satisfactory results.

Evaporation

The rate of evaporation in hot, dry climates can be excessive during the summer period. The best solution is to irrigate during the night. Frequent irrigations and the use of sprinklers which produce large drop sizes will also help; providing the soil and crop can withstand the treatment.

5

Pest and disease control

The 20th century saw the development of a multitude of chemicals such as fertilisers, pesticides and growth stimulants to aid the modern farmer. These have greatly increased overall yields, but unfortunately have created new problems, including:

- Reliance of farmers on expensive chemicals
- Health concerns regarding the use of such chemicals
- Contamination of the environment

Pests and diseases need not be seen as a problem if the numbers or extent of infestation is fairly low and the amount of damage is minor. For many crops, particularly those grown by organic means, some small level of pest and disease damage may be acceptable. However, pest and disease damage/infestation can render some produce virtually worthless, as it cannot be sold at a fair price.

It would be difficult for most farmers to completely stop using chemicals but it is possible to significantly reduce dependence on them. Nature has provided a whole range of methods to control pests and diseases without the side effects that artificial chemicals have. The main point to remember about natural control methods is that they rarely achieve the same degree of control as some conventional chemical methods.

Pest management and systems thinking

One important concept of sustainable agriculture is that it takes a holistic approach to the entire farm system. Instead of conventional agricultural practices where, for example, a specific chemical is used to almost completely eradicate a certain pest or disease, sustainable agriculture dictates a much broader-based approach. Cropping systems may be designed so that rotation reduces the need for chemical spraying. Chemical spraying may also be minimised by choosing disease-resistant varieties and by growing crops in areas where pests find it difficult to survive on that crop. The downstream effects and total system cost of that chemical spray are taken into account. Certain levels of pests and diseases may be acceptable in a sustainable system as their overall cost and effect on the operation may be less than the chemicals or other conventional methods used to control them.

Integrated pest management

Chemicals kill pests and diseases effectively, but there can be problems if you don't use the right chemical or the right method. There are of course other ways to control pests and diseases, but other methods rarely give the same degree of control as chemicals. Nevertheless, the preferred option these days is usually to use a combination of control techniques. The concept is that:

- Nothing should be used to the detriment of the environment or to the extent that pests get accustomed to the method (as they may build up resistance).
- Each different technique weakens the pest or disease that little bit more, the overall effect is cumulative, and may be quite effective.
- Expensive controls (eg some costly chemicals) are used in limited quantities, keeping costs lower.

This idea of using a combination of different control techniques which each contribute to the overall control is known as 'Integrated Pest Management' or 'IPM'. The principle of IPM relies upon creating, as far as possible, an environment where there is a balance between sustainable environmental practices and profitable farming. For example, birds can be of benefit when they eat insect pests, but they can also become pests themselves.

The basis for natural control is to regulate the environment to give nature the very best chance of keeping pest and disease problems in check. This may take some time to achieve and result initially in high losses of plants and animals until a balance is found. This is also very difficult to achieve for the farmer surrounded by other farms where other means of pests and disease control are used.

Integrated pest management allows the use of pesticides and herbicides, but only as part of an overall management program.

Biointensive integrated pest management

Biointensive integrated pest management is a variation of conventional integrated pest management. Conventional IPM has been criticised for using pesticides as a first resort when other methods of pest management are not successful. Biointensive IPM really emphasises the importance of understanding the ecological basis of pest infestations.

Biointensive IPM asks the following questions:

- Why is the pest there and how did it arrive?
- Why don't the natural predators control the pest?

Proponents of biointensive IPM claim that it will decrease the chemical use and costs of conventional IPM. Biointensive IPM requires that the agricultural system be redesigned to favour pest predators and to actively disadvantage pests. For example, whilst integrated pest management is used in monocultures, biointensive IPM would require the system to be redesigned to perhaps incorporate a less pest friendly system such as a polyculture.

Pesticides: a vicious cycle

There is an intrinsic problem with many pesticides, and that is that they wipe out almost every individual in a pest population – except any that happen to be naturally resistant to the pesticide. The few individuals that survive pesticide application are then able to have offspring, and so pass on their resistance to the pesticide.

With the rest of their population gone, these individuals tend to multiply rapidly. The only difference is that this time, the population is almost entirely resistant to the pesticide.

Most pesticides kill off natural enemies along with the pests. So, with their predators gone, pest populations can explode to a much higher level. In addition, some potential pests that are normally kept in check by natural enemies become a real problem after a pesticide wipes out their predators.

Only a small amount of any pesticide actually contacts the target species. At best, the remainder may break down – or it may be carried by wind, water and soil to kill of non target organisms, and to be taken in by higher predators, even reaching humans in the food we consume.

Controlling pests and diseases in plants

There are a number of different things that can be done to help to control pests and diseases within a sustainable agricultural system. The following will be discussed in detail below:

1. Cultural controls – the methods used to grow plants
2. Physical controls – the methods which physically interfere with pests or diseases
3. Sprays and dusts – natural products which control pests or diseases, some of them do so without undesirable side effects
4. Biological controls – where other organisms control the pest or disease, by such means as directly attacking the problem, by repelling it, or by attracting or luring pests to a place where they can be easily trapped or collected, and then destroyed or removed elsewhere
5. Companion planting – plants growing near one another can enhance or inhibit each other's growth and vigour
6. Legislation – Government laws, for example those covering quarantine can help to address a pest/disease problem
7. Genetic engineering – plants are now being bred that are genetically resistant to certain pests and diseases

Cultural controls

Growing your plants at the correct time of year in a position that suits them will reduce the likelihood of pest and disease problems occurring. Growing plants in poor conditions leaves them open to pest and disease problems.

Choose healthy plants

As a general rule, healthy plants will show greater resistance to pest and disease attack and will be more likely to recover if they are attacked. When propagating your own plants, make sure you only use propagating material from healthy vigorous parent plants. When buying plants, make sure you only choose healthy-looking ones. Carefully inspect the plant, the surface of the potting mix, the plant container and associated equipment for signs of pest and disease infestation. When selecting bare-rooted plants, such as fruit trees, carefully check the roots for signs of damage, abnormal swellings or growths, etc. A little time and care taken in the selection of your plants will usually mean a big reduction in pest and disease problems later on.

Choose resistant plant varieties

Some plants seem to have few pest and disease problems. These plants are said to be resistant or tolerant. In some cases this is because pests and diseases are simply not attracted to that particular type of plant. In other cases the plant directly affects any insects or pests attempting to live on it, for example by exuding chemicals that repel the pest. Some plants also have a greater ability to withstand insect or disease damage than others. In some cases the regular pruning back of foliage by insects can help to keep plants vigorous. By choosing such resistant or tolerant plants you will reduce the likelihood of problems occurring.

Crop rotation

Different crops will attract different pest and disease problems. It is always a good idea to grow crops on a rotation system, as growing the same type of plant in the same soil year after year can produce ideal breeding conditions for certain types of pests and diseases. By changing the crops around, the host plants are always different. This prevents any buildup and such problems are not carried over from year to year. Crop rotation plays a particularly vital role in controlling root diseases. A lot of crops will also have different nutritional needs so, again, rotation will prevent specific crop nutrients in the soil from becoming exhausted.

Timed planting

Although it is not always possible, some crops can be grown at the time of year when populations of the pests or diseases that affect them are at their lowest. Cabbages and cauliflower, for example, are less affected by the caterpillars of the cabbage white butterfly if they are grown through winter, avoiding warmer seasons when the butterfly is common. Growing crops early on in their normal growing season before pests and diseases have a chance to build up can also help reduce pest and disease problems. In this case you can start vegetables, flowers, etc off early, in a glasshouse or cold frame, so that they can be established outside as early as possible.

Irrigation

In hot weather, too much water on the surface of the ground or the leaves will encourage fungal diseases and some insects. By using drip irrigation, these problems can be decreased. Flood irrigating an area occasionally can be used to drown some pests that live in the soil

Increasing plant diversity

A farm which produces a greater variety of plants has far less chance of suffering serious losses as a result of a pest or disease plague. This needs to be balanced against having the expertise and equipment required to produce a greater variety of crops.

Mulching

At harvest time, crops that have contact with the soil, such as strawberries, marrows and zucchinis are very susceptible to fungal diseases. Mulch can be laid under the plants so that the crop does not come into direct contact with the soil. Mulch is usually a material, such as straw, which helps to keep the crop clean and dry and therefore reduces the instances of fungal disease.

Cleanliness and hygiene

If the area around plants is kept clean and free of pests and disease, there is less chance of the plants being affected. This can be achieved in such ways as:

- Being careful that you do not import soil that may be full of pests and diseases, weed seeds or other problems
- When buying plants, be careful that the soils and potting mixes that they are grown in don't have similar problems; where possible obtain your soil and plants from a reputable supplier
- Where possible, don't leave plants or plant parts affected by pests or diseases near healthy plants.
- Use clean, sharp tools when working with plants; regularly dip or rub over tools such as secateurs, handsaws and knives with an antiseptic such as methylated spirits or Dettol. Keep cutting tools sharp to prevent tearing or ripping of plant material, which may make the plant more susceptible to attack.
- Ensure that any machinery (eg planters, harvesters) used where pests, diseases or weeds are a problem are thoroughly cleaned before being used elsewhere.
- Keep a close eye on plants and do something about problems immediately they are noticed.
- Avoid having muddy areas, if possible, improve the drainage so drains take excess water away, and not just redistribute it elsewhere, including any diseases that might be in it.
- Clear away any weeds near your crops. Many weeds will act as hosts to diseases and pests which affect your crops. Keep in mind that some pests, like aphids and leafhoppers, also spread viral and bacterial diseases as they suck sap and move from plant to plant.

Climate modification

Climate modification can be used to reduce pest or disease populations by creating environments they don't like. For example:

- improving ventilation will often help control fungal problems
- shady conditions may promote fungal and other diseases or weaken a plant, causing it to be more susceptible to attack

- creating drier or damper conditions will often deter different pests; for example some ants don't like very moist soil
- overhead spray irrigation fosters disease more than does flood or trickle irrigation

Physical controls

Hand removal

Many pests can be simply removed by hand, especially if you catch the problem before it spreads too far. Snails on a rainy night can be squashed or otherwise killed. Small infestations of caterpillars or grasshoppers can be squashed between your fingers (preferably while you are wearing gloves) or knocked to the ground and squashed with your feet. Leaves with fungal problems or insect problems such as scale can be picked off and burnt. Obviously, this method is only applicable to small areas.

Pruning

Pruning can be used in two ways to control pests and diseases. The first is by modifying the shape of the plant in a way that makes it less likely to attack; for example, by removing damaged or rubbing branches, by opening up the centre of the plant to improve air circulation, or by removing areas that could provide shelter for pests. The second way is by directly pruning away plant material already affected by pests and diseases to prevent the spread of such problems. The pruned material should ideally be burnt.

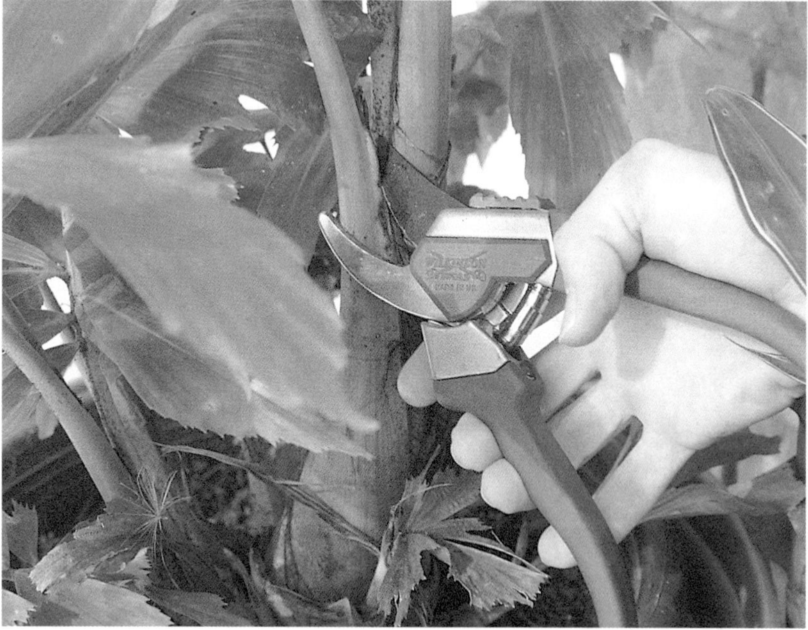

Figure 5.1 Pruning can reduce the impact of pests and diseases in plants. Pruning removes diseased branches and can open the plant up to increase air circulation. A poorly structured plant is more prone to sickness.

Hosing

Some pests can be readily knocked off plants by simply hosing them with a strong spray of water. This is particularly useful for non-flying pests.

Physical barriers

Physical barriers can often be a very effective means of preventing pests and diseases from reaching your plants. Examples of physical barriers are:

- Fences or tree guards to prevent rabbits, dogs, grazing animals, children, etc. from coming into contact with plants
- Netting to keep birds and fruit bats away from fruit
- Greenhouses to isolate plants from insects, fungal spores, etc
- Metal collars on trees to stop possums or crawling insects climbing up them
- Insect screens on small cages over vegetables, or to cover a ventilation opening on a greenhouse

Traps

Traps work by catching pests, either some distance from the crop or where the pest normally occurs. The trap may consist of a sticky substance placed in the path of a pest so that it becomes stuck or it may be a container with a lure inside which will attract the pest inside, where it can be collected later.

Yellow attracts many insects (eg aphids). Yellow cards covered with something sticky (eg honey) will attract insects which are then stuck to the card surface. The cards are periodically collected, burnt and replaced with new ones. Other colours will attract different pests, eg white will attract thrip.

Tin foil hung near your plants will reflect light. This can confuse aphids and reduce attack. Foil can also provide extra light and warmth early in the season.

Table 9 Some specific methods for controlling certain insect pests

Ants	Sticky substances such as bituminous paint on the base of a tree trunk to stop ants crawling up into branches
Aphids	A piece of cardboard painted yellow and coated with a thin layer of honey; when aphids become stuck, remove and burn
Bugs	A ring of camphor around a tree will repel bugs and some other insects
Caterpillars	Caterpillars are the larvae of moths and butterflies, which can be attracted into light traps, thus reducing caterpillar populations. Most insects generally avoid red, yellow or orange lights but white, green or pink fluorescent lights can be effective. Some commercial light traps incorporate an electric grid which kills insects attracted to the light. These traps need cleaning out regularly to work well.
Codling Moth	Wrap several layers of corrugated cardboard around a tree trunk in early spring to attract moth larvae; remove in summer with the larvae attached and burn
Whitefly	A yellow card covered with a thin layer of honey, works best early in the season

Repellent devices

There are a variety of devices that can be used to repel pests, in particular birds. The devices usually need to be moved around to work, and may only work for a short time, so only use them at the most critical times.

Scarecrows

Scarecrows will often work for a short time. Birds and other animals become cautious when there is something different in an area they have been visiting. When a scarecrow first appears, birds will sometimes avoid that area for a few days (or perhaps weeks) until they become used to it. This is the way to use a scarecrow:

- Put it up just as your fruit is starting to ripen
- Change its position every few days
- Change its appearance if you can (eg change clothing)
- Have parts of the clothing loose so they will blow in the wind and create movement

Figure 5.2 Scarecrow used for bird control in an organic garden.

Scare guns

These are commonly used to protect fruit and vegetable crops, but generally only on commercial scale crops. A major problem with these devices is that if your crop is near your house, the noise from the gun can be very annoying.

Repelling dogs, possums, mice, rats etc

Animals with a keen sense of smell are often deterred by a change in the smell of an area. Camphor, pepper, peppermint oil and other such things are often used to discourage these animals or to break their habit of visiting an area.

Repelling cabbage moth

The female (which lays eggs) is repelled by the smell of tar. You might put out some fresh bituminous paint, tree wound paint or use builder's tar paper to put a collar around each cabbage plant.

Bird wires

These wires hum in the wind and can scare birds away.

Hawk silhouette

A model or cut-out of a hawk shape is fixed to a post or overhead on a wire. This scares away fruit and nut-eating birds.

Sprays and dusts

Some chemical sprays and dusts can be used within a sustainable framework, although obviously the less used, the better. In many cases the amount of chemical used to control pests and diseases can be significantly reduced by:

- Correct identification of the problem
- Correct timing of applications – ensuring that the chemical is supplied at the time when it will be most effective
- Using the most efficient application methods – thereby minimising the amount of chemical required, and ensuring it gets most effectively to where the problem is
- Careful selection of the chemical/s to be used

Using sprays and dusts

The action of sprays and dusts is variable. Systemic pesticides act by being absorbed by the plant, whereas contact pesticides only work by direct contact with the problem. Selective pesticides only kill the target organism, while non-selective pesticides may kill desirable organisms. Some pesticides are residual and remain in the soil after use.

Always use the least hazardous chemical available and always follow the application rates and safety instructions on the label.

Organic sprays and dusts

A number of organically sound sprays and dusts are available. In most countries, organic gardeners can legally use certain chemicals that are derived from botanical and mineral-bearing sources. Whilst these chemicals may be toxic, they break down more rapidly than other non organic chemicals. They may not have a 'bulldozer effect' like some of the potent chemicals, but they are safer for both you and the environment and, if used properly, will keep many problems well under control. Ideally, they can be incorporated into an integrated pest management system.

Pesticides

Pyrethrum

Pyrethrum is a naturally occurring plant extract that is used widely in sprays and aerosols for home use. It has also been used successfully in broader scale agricultural production. Synthetic pyrethroid sprays are now widely used in agricultural and horticultural production (eg Permethrin). To be effective, the spray must make contact with the insect pest. It is toxic to fish.

Bacillus thuringiensis

Sold as Dipel®, it is a naturally occurring bacteria supplied in a powder-like form. This selective control is used on a wide variety of caterpillars.

Nicotine

Extracted from the tobacco plant, it kills sap sucking insects. It is non-systemic and non-residual but toxic to mammals if swallowed.

Rotenone

Also known as Derris Dust, Rotenone is an extract from a plant root. It is non-selective, non-systemic and has a low level of persistence in the environment although it is toxic to fish. It is often mixed with sulphur compounds.

Hydrocarbon oils

Hydrocarbon oils include white oil, made from paraffin, and winter oil, a petroleum-based product. They are used as a contact pesticide for sap-sucking insects such as scale and mites. They may damage the foliage of some plants, particularly if the plant is in direct sunlight.

Sulphur

Sulphur is toxic to mites and scale, although it is more commonly used as a fungicide. It is not suitable for use in hot weather.

Sulphur-based products are registered and approved as organic fungicides. Sulphur comes in a variety of forms including wettable sulphur, lime-sulphur and sulphur dust. It acts to protect leaves with a chemical coating and can also control established fungal infections such as powdery mildew.

Neem

Neem is a plant extract that interferes with insects' hormone system, preventing reproduction. Whilst not yet widely available, it offers great potential as a non-selective, non-residual insecticide.

Sabadilla

Sabadilla is an insecticide made from the ripe seeds of the South American sabadilla lily (*Schoenocaulon officinale*). It contains an alkaloid known as veratrine. Sabadilla is among the least toxic of botanical insecticides, and it breaks down rapidly in sunlight. It is marketed under the trade names Red Devil or Natural Guard. Sabadilla is effective against caterpillars, leaf hoppers, thrips, stink bugs and squash bugs.

Fungicides

Fungal diseases are more difficult to control than insect pests. The organic and synthetic sprays and dusts described below will not control all fungal pests.

Copper

Copper is sold in a variety of forms, including Bordeaux (a mixture of copper sulphate and lime) and copper oxychloride. Some copper-based fungicides are registered and approved for use in organic farming. It is used on a wide variety of vines, fruit trees and other plants and is usually applied to leaves and/or stems before infection is likely to occur. This provides a protective chemical barrier for both fungal and bacterial diseases. Copper can, however, cause long-term soil toxicity problems. Copper compounds can also damage the leaves of some plants and withholding periods should be observed before harvest.

Potassium permanganate

Also known as Condy's crystals, it is used as a spray for control of powdery mildew.

Teas

Various other products can be used to make non-residual organic fungicide sprays. They include garlic, chives, horseradish, nettles and milk.

Chemical control of pests and diseases in plants

There may be occasions when a particularly resilient pest or disease cannot be controlled by organic methods. In this situation it may be necessary to use a synthetic chemical control to bring the problem to a manageable level, when more sustainable methods can be put into practice.

Figure 5.3 Boom sprayer on tractor. Whilst sustainable agriculturalists aim to reduce the use of chemical pesticides and herbicides, there are times when chemicals need to be used in small quantities.

Advantages of chemical control

- Reliable
- Low labour costs
- Covers broad areas
- Quick results

Disadvantages of chemical controls

- May kill non target organisms, including desirable species
- Sprayed areas are vulnerable to new pest infestations
- Loss of status as an organic farm
- May leave poisonous residue in the soil

Synthetic sprays and dusts

Many of these compounds are highly toxic and, if used incorrectly, can harm humans, livestock, soil and waterways. Always read the label before use.

Insecticides

Carbamates

Including one plant extract, these compounds act both systemically and on contact. Most are non-selective and work by interfering with the nervous system. Two of the most common carbamates are Carbaryl (toxic to most insects, including bees) and Methiocarb (for snails, slugs and slaters). They are less persistent in the environment than the chemical groups described below.

Organo-phosphates
This large group of pesticides has varying levels of toxicity, which work by interfering with the nervous system. They can act systemically or by direct contact with the pest, and most are residual. Examples include the non-selective compounds Dimethoate (Rogor®) and Malathion (Maldison®). Overuse has led to many pests becoming resistant to these chemicals.

Organo-chlorines
These are highly toxic, non-selective, residual pesticides, and include DDT and dieldrin. Most of the chemicals in this group have now been banned or are severely restricted in most parts of the world.

Metal-based compounds
These include chemicals based upon heavy metals such as lead and mercury. These chemicals are highly dangerous and are now banned in most parts of the world.

Fungicides

Carbamates
These include Zineb, used as a chemical protection, and Dithane (Mancozeb®) which acts on contact.

Systemic fungicides
In most cases the entire plant must be covered for these chemicals to be effective. Benomyl (Benlate®) is one such fungicide which is a wettable powder used to control mildew, rots and mould.

Soil fungicides
These are residual chemicals applied to prevent fungal diseases developing and include furalaxyl (Fongarid®), used for treating pythium and phytophthora.

Chemical application techniques

Where chemicals are used, it is important to minimise their use. One way to do this is to ensure they are applied efficiently. Appropriate nozzles should be used and the correct nozzle pressure will reduce wastage. Misdirected sprays delivered by the wrong nozzle will increase chemical wastage (and cost) and the environmental impact of the operation.

Where pest populations are isolated, spot treatments are best. Monitoring of pest populations will tell a manager whether broadscale treatment is really necessary.

In minimal cultivation systems, herbicides can be applied in bands. This puts the herbicide only where it is needed – usually in soil that has been disturbed by tillage or seed planting where weeds are most likely to occur.

Biological controls

Biological control is the use of a biologically derived agent (ie plant, insect, animal) to control pests and diseases. This commonly involves the use of diseases which affect the pest or weed (the disease might be spread by an insect) or beneficial insects which either eat or parasitise the pest. These control agents are sometimes known as antagonistic organisms.

Antagonistic organisms

A balance usually develops in nature among organisms, both plant and animal. Certain organisms are antagonistic to others and retard their growth. Environmental or human-induced changes that upset this balance by eliminating one of the organisms can lead to explosive proliferation of the others and to subsequent attacks on vulnerable crop plants. Biological control in such a situation would consist of introducing the antagonising organism of the pest into the area, thus bringing it under control again. Olive parlatoria scale threatened the existence of the California olive industry, but two parasitic wasps introduced from Asia became well established and practically eliminated the scale.

There are three main approaches to biocontrol. These are:

1. The introduction of parasites and predators, where natural enemies are introduced to control exotic pests, as in the case of cottony cushion scale, which was introduced to California from overseas without its natural predators. The importation of vedalia beetles from Australia virtually eradicated this pest very quickly, and keeps it in check to this day.
2. Conservation of existing natural enemies by changing spraying programs (we can't always just stop spraying; we need to build up the natural enemies to a useful level first). This can be achieved by using selective chemicals or by changing when we spray, as some insects are active at different times of the day. We can also reduce the rates of the chemicals that we use. Another method of conserving natural enemies is to change the way in which plants are cropped. This can be done by such methods as staggering planting times to reduce the impact of having a crop all at one stage when it may be more prone to attack or infestation; by the use of companion plants; by increasing crop diversity, by mixing crop species, and by maintaining groundcover in orchards to provide habitats for beneficieal parasites.
3. New natural enemies can be developed by scientists growing larger numbers of predators or parasites, or by adding additional numbers of natural enemies collected or purchased from elsewhere. Producing and marketing biological control agents has now become a major business in Europe and the USA, with small-scale activity also in Australia.

Other approaches to biocontrol that are being actively researched are the development of plants with increased resistance to pests and diseases; the use of natural chemicals such as hormones or sex scents to either attract (to a trap or away from plants), repel or kill these types of problems; the use of sterile insects to upset reproductive cycles, and the use of plant derivatives such as pyrethrum as pesticides.

Advantages of biocontrol methods

1. In contrast to many chemicals, antagonistic organisms don't damage plants
2. No residues are left, as in the case of many chemicals
3. You don't have to wait (ie there is no witholding period) before harvesting produce, as commonly occurs when using chemicals
4. It's less costly than using chemicals and, unlike chemicals where repeat applications are generally necessary, predators and parasites may offer continuous control as they continue to breed
5. These organisms can spread, often very rapidly, controlling pests and diseases over large areas
6. Pests and diseases are unlikely to build up resistance to these organisms, as is often the case when using chemicals
7. These organisms are generally predators or parasites of specific pests or diseases and will not affect other organisms

Disadvantages of biocontrol methods

1. They are often very slow acting in comparison to chemicals
2. The degree of control is often not as high as with chemicals
3. It is often very hard to find predators or parasites of some pests, especially ones that are specific to that pest or disease, rather than a number of organisms
4. The mobility of antagonistic organisms can sometimes be a disadvantage. What may be a pest or disease in one area may not be one elsewhere; for example blackberries are a declared noxious weed in some areas of Australia, but are also grown commercially for their berries. The recent introduction to Australia of a blackberry rust as a means of blackberry control may potentially affect crop varieties.

The advantages certainly far outweigh the disadvantages in the long term if not in the short term, particularly in terms of the effects on the environment.

Predators

These include a wide range of animals such as lizards, frogs dragonflies, spiders and birds. To be effective they need places to shelter and breed (eg hollow logs), food (insects, nectar, pollen) and water. Insect-eating birds can be attracted into the area by providing plantations of native plants like gums, grevilleas and bottlebrushes. Many insects are also good predators of pests:

- Ladybird beetles and their larvae eat aphids
- Hover flies (Syrphid flies) eat aphids
- Lacewing will control mites, caterpillars, aphids, thrips, mealy bugs and some scales
- Predatory mites eat other pest mites. They can be purchased and released in the crop.
- Praying mantis eat most other insects – pests or not

There are many other predators and as long as there is a suitable environment for them and the sprays are minimised, they can do much of the work for you.

Attracting parasites

Wasps attack many types of insects including caterpillars. Some plants (eg chamomile, celery, hyssop, tansy, dill, and yarrow) can be planted to attract such wasps to the garden.

- Woolly aphids parasites are attracted by clover (*Trifolium* sp.)
- Lacewings which feed on aphis and other insects are attracted by sunflowers
- Goldenrod (*Solidago* sp.) attracts preying mantis and some other predators
- Hoverflies are strongly attracted to buckwheat

The Rodale Institute Research Centre in America tested a range of plants that attract beneficial insects (reported in *Organic Gardening* May/June 1991). The most effective plant was tansy, followed by caraway. Dill, white cosmos and buckwheat were also good. Fennel was effective, but is not recommended in Australia where it is a weed. The other plants could also be a problem in some areas.

Trap or decoy plants

These work by attracting insects away from your desired plants. Decoy plants should be grown away from the main crop, otherwise you may just increase the population of pests, some of which may move into your crop, especially if the decoy plants are cut down or not irrigated.

Table 11 Decoy plants

Plant	Attracted insects
Datura	Chewing beetles
Hibiscus	Harlequin bugs
Hyssop	Cabbage white butterfly
Mustard	Cabbage butterflies
Nasturtium	Aphids

Pheromone traps

Female insects, such as fruit flies and codling moths give off a scent called a pheromone, which attracts males for reproduction. Traps have been devised using these pheromones as an attractant. These traps are used by scientists to see how many insects are present in one area compared to another, and whether the numbers of insects are increasing. Pheromone traps are available commercially as a control method for specific pests such as fruit fly.

Beneficial plants

Some plants have beneficial effects on the health or growth of other plants growing nearby. The obvious example is legumes (ie pea type plants) which have symbiotic relationships with a variety of soil micro-organisms that result in the extraction of nitrogen from the air, converting it into solid nodules on the plant roots where it eventually becomes available for use by other plants growing alongside them.

Table 10 Other beneficial plants

Plant	Reported effect
Alyssum	attracts hoverflies which eat aphids
Anise (*Pimpinella anisum*)	deters aphis and cabbage grubs
Basil	repels flies
Borage	attracts bees
Buckwheat	attracts beneficial insects
Calendula	deters asparagus beetle
Caraway	attracts beneficial insects
Carrot	deters onion fly
Celery	deters cabbage butterfly
Chrysanthemum coccineus	reduces nematodes
Clover	fixes nitrogen and attracts beneficial insects
Coleus caninus	deters dogs and cats
Cosmos *'White Sensation'*	attracts beneficial insects
Dandelion	repels chewing beetles
Dill	repels aphis and red spider and attracts beneficial insects
Fennel	repels fleas and aphis
Garlic	general insect repellent, deters Japanese beetles, also having high levels of sulphur it is a mild fungicide
Horseradish	deters chewing beetles
Lavender	attracts bees
Leek	deters carrot fly
Lucerne	fixes nitrogen and attracts beneficials
Marigold (*Tagetes* sp.)	exudation from roots reduces nematodes, attracts hoverflies, which eat aphids and parasitic wasps
Onion	deters carrot fly and chewing beetles
Peppermint	deters many insects (particularly ants) and rodents
Petunia	repels bean beetles and some bugs
Rosemary	deters bean beetle, carrot fly, cabbage grubs, attracts bees
Rue	deters beetles and fleas
Santolina	deters corn grubs
Southernwood	deters cabbage moths, carrot flies
Tansy	deters ants, flies, mosquitoes and attracts beneficial insects
Thyme	deters caterpillars and whitefly and attracts bees, improving pollination
Wormwood	deters many insects, slugs and snails

Companion planting

Companion planting is based on the idea that certain plants grow better in close proximity to other plants. For example, garlic planted under a peach tree will improve the performance of the peach by deterring the development of peach leaf curl, a disease which is very common in peaches. Many of the claimed effects of companion planting have not as yet been scientifically proven. Research has been undertaken which disputes some claims and proves the validity of others. Some companion planting effects are certain and very strong. Others are mild or even questionable.

Although readily used by the home gardener, companion planting principles can be used by the farmer on a larger scale.

You should treat the recommendations for companion planting as combinations which can be tried out but, unless stated otherwise, do not expect dramatic results. Companion planting must always be given time to work. The effects are rarely immediate.

Companion planting works by having one or several effects such as the ones described below.

Repellent plants

Certain plants will repel insects or other pests from an area. This usually works because of the aroma being released from the plant. Plants said to work in this way are:

- Fennel for fleas
- Peppermint for mice and rats
- Wormwood for snakes
- Pennyroyal for ants
- Tansy for flies

Attractant plants

These are plants which attract pests away from valuable plants (ie 'sacrifice plants') or which attract predators which, in turn, control pests. For example:

- Clover may attract woolly aphis away from an apple tree
- Moths are attracted to some types of lavender
- Hyssop attracts cabbage white butterfly
- Marshmallow attracts harlequin bugs
- Lacewings and predatory mites are attracted to sunflowers

Plants which affect the soil

Plants can affect the soil in many different ways to create desirable or undesirable effects for other plants. For example:

- Legumes such as peas, beans or lupins have colonies of bacteria on their roots which have the ability to take nitrogen from the air and convert it into a form of nitrogen which the plant can absorb
- French marigolds exude a chemical from their roots which deters the development of nematodes in the soil
- Garlic or other onion type plants will increase the level of sulphur in the soil in desirable forms, leading to some control over fungal diseases
- Some plants accumulate certain nutrients from the soil better than others; when these plants die and are composted, they supply those nutrients back to the soil in a more 'available' form. (NB: You may need to compost the entire plant, including roots.)

Table 12 Nutrient accumulating plants

Nutrient	Plants which accumulate it
Nitrogen	All legumes (eg clover, peas, beans, wattles, lucerne); most plants when young are higher in nitrogen
Phosphorus	Dock, buttercup, comfrey, oak leaves, yarrow
Potassium	Dock, buttercup, comfrey, coltsfoot, couch grass, maple leaves, stinging nettle, sunflower, tansy, tobacco, vetch, yarrow
Calcium	Beech (*Fagus* sp.), brassicas, buckwheat, comfrey, dandelion, melons, oak (*Quercus* sp.), Robinia, stinging nettle (*Urtica* sp.)
Magnesium	Beech (*Fagus* sp.), chicory, coltsfoot, equisetum, oak, potato, salad burnet, *Potentilla anserina* and yarrow
Iron	Beans, buttercup, chickweed, chicory, coltsfoot, comfrey, dandelion, foxglove, *Potentilla anserina* and stinging nettle
Manganese	Silver beet, spinach, comfrey, buttercup
Sulphur	Allium (eg garlic, onion, chives), brassicas, coltsfoot, fat hen
Copper	Chickweed, dandelion, plantain, stinging nettle, vetch, yarrow
Boron	Cabbage, cauliflower, apple, euphorbia
Cobolt	Buttercup, comfrey, equisetum, willow herb

NB: The information in Table 12 can also be used as a guide to what plant materials are a source of different nutrients. For example, if you have a manganese deficiency, you might try applying quantities of compost made from silver beet, spinach or comfrey, which all accumulate that nutrient.

Table 13 Companion plants

Plant	Companion	Comments
Apple	Nasturtium	Deters aphids in apple tree
Apple	Parsnips	Parsnip flowers attract parasites which kill codling moth
Asparagus	Tomatoes	A chemical called 'solanine' in tomatoes will deter beetles on asparagus
Basil	Tomatoes	Deters both diseases and pests in tomatoes
Bean	Summer savory	Deters pests and improves growth
Bean	Carrots	Carrots improve the growth of beans
Beet (red or silver)	Beans	Beans improve growth of beets
Cabbage	Rosemary, hyssop, wormwood or southernwood	These plants all deter cabbage moth
Cabbage	Roman chamomile	Improves growth and flavour of cabbage
Cabbage	Tomato	Tomatoes help deter cabbage grubs
Celery	Leeks, tomatoes, cabbage and cauliflower	
Citrus	Oak or Ficus	Some authorities suggest oaks and figs give off a chemical which has a protective action on citrus
Cucumber	Bush bean	Beans enrich the soil with nitrogen which cucumbers use in large amounts
Cucurbits	Nasturtium	Deters aphids and other pests
Fruit trees	Garlic or chives	Planted around base will deter insects climbing up into the tree

Table 13 Companion plants (continued)

Plant	Companion	Comments
Fruit	Catnip	A couple of plants will attract cats to the area, in turn this deters birds from attacking the fruit.
Grape	Mulberry	Grapes allowed to climb in a mulberry tree are less susceptible to disease (though they are more difficult to pick)
Lettuce	Radish	Radish grown with lettuce during warm weather make the lettuce crop more succulent
Lettuce	Cucumber, carrots, strawberries and onions	
Onion	Roman chamomile	Improves flavour and growth of onions
Peach tree	Garlic	Deters development of peach leaf curl
Peas	Carrots	Carrot roots give off a chemical which benefits peas
Pumpkin	Nasturtium	Helps protect pumpkin from various insects
Potato	Beans, horseradish, cabbage or corn	Plant any of these at the edge of a potato plot
Radish	Chervil	Chervil beside radish makes radish taste hotter
Radish	Climbing beans or peas	These make radish grow stronger
Rose	Garlic	Garlic deters beetles, aphids and fungal diseases on roses such as black spot
Rose	Parsley, mignonette lettuce	
Shrubs	Beech (*Fagus* spp.)	In medium to large gardens, the beech improves soil fertility
Tomato	Bee balm	Improves flavour of fruit
Tomato	Borage	Attracts bees, improves tomato flavour, deters tomato worm
Tomato	Basil	Improves growth and flavour
Tomato	French marigold	Marigolds deter white fly on tomatoes
Trees	Garlic	Garlic planted in a circle under trees deters borers
Trees	Birch (*Betula* spp.)	Birch roots secrete a chemical which speeds decomposition of leaf litter or any other organic material on the ground
Vegetables	Mustard or marigold	Both mustard and marigolds deter nematodes; many vegetables are susceptible to nematode problems, including beans, cucurbits, celery, lettuce, capsicum, eggplant, tomato and okra
Vegetables	Marjoram	Marjoram can improve the flavour of many different vegetables

Table 14 Herb repellents

To repel	You should plant
Rodents (mice and rats)	Mints (peppermint is best), catnip, daffodil, jonquil, grape hyacinth and scilla
Dogs and cats	Coleus *caninus*
Aphids	Lad's love (southernwood)
Cabbage moth	Lad's love, chamomile, sage
Flies	Basil, rue, mints, tansy
Insects (in general)	Basil, chamomile, pennyroyal, mints and tansy (most plants in the Lamiaceae family)

Table 15 Resistant plants

To resist	Plant	Avoid
Cinnamon fungus	Westringia	Prostanthera
Scale insects	Grevilleas	Leptospermum, lemon, almonds and eriostemon
Wood rots	Oak	Elm
Air Pollution	*Melaleuca wilsonii, Acacia longifolia, Eucalyptus maculata, Platanus hybridus*	*Melaleuca ericifolia, Acacia howitii, Eucalyptus nicholii*

Table 16 Bait plants

To attract	You should plant
Dogs and cats	Anise, catnip (cats only)
Birds	Banksia, Grevillea, Cotoneaster, Cretagus, Hakea, Pyracantha
Insects	Onion, garlic, chives, nasturtium
Bees	Bee balm
Cabbage moth	Hyssop

The plants listed in Table 17 are said to control pests as listed. You should remember that companion planting controls are unlikely to be total. Pest numbers are normally reduced, not eliminated.

Table 17 Pest control plants

Insect	Plants
Aphids	Nasturtium, spearmint, stinging nettle, southernwood, garlic
Cabbage butterfly	Sage, rosemary, hyssop, thyme, mint, wormwood, southernwood
Flies	Rue, tansy, nut trees
Slugs	Oak leaf mulch, tan bark, woodshavings
Moths	Sage, santolina, lavender, mint, stinging nettle
Nematodes	Marigolds (*Tagetes* spp.), *Chrysanthemum coccineum*
Weevils	Garlic

Do not plant the following combinations

- Beans with onions, garlic, leeks or chives
- Broccoli, cabbage, cauliflower and other Brassicas with tomatoes, beans or strawberries
- Potatoes with pumpkin, tomato, raspberry, squash, cucumber or sunflower (these stunt the growth and disease resistance of potatoes)
- Tomatoes with apricot trees as they give off a chemical which inhibits growth
- Sunflowers or Jerusalem artichoke with potatoes, as they slow each other's growth
- Roses with Buxus (eg English box), as Buxus roots spread and compete strongly with the rose roots

Legislation

Chemical use

In most parts of the world, the laws and regulations for agricultural chemical use are the responsibility of regional and sometimes national governments. Governments pass acts of parliament that give broad guidelines for the standards to be met regarding issues such as dangerous goods, aerial spraying and occupational health and safety. Specific guidelines for the handling, transport, storage and use of agricultural chemicals are explained in detail in supporting regulations.

The laws and regulations covering agricultural chemicals are revised periodically. You should always first check with your local authorities to ensure relevant chemicals are currently licensed for use in your area. If any proscribed (banned) chemicals are present on your farm, they should be handed to the appropriate authorities for safe disposal.

Quarantine

Because they have no natural predators, exotic pests have the potential to decimate agricultural and horticultural crops. The problem can take many forms:

- Small pests that 'hitchhike' between countries, for example the fire ant is believed to have been transferred to Australia in a shipment of fruit
- Fungal spores are very small and can be easily transported on plant material – even in soil on the soles of shoes
- Weeds and their seeds can hitchhike or come in with other plant material
- Aquatic pests come in the ballast of ships, this has become a problem in waterways around the world

National governments have passed laws that restrict the movement of animals and plant material from country to country. Prohibited imports include some animals, seeds, plants and even wooden souvenirs brought by tourists.

To restrict the spread of pests or diseases in regional areas suffering pest or disease, the movement of certain plants, fruits, vegetables, etc., are restricted within or between certain boundaries. It is extremely important that these rules or laws are complied with. Advice on local regulations may be found at local council offices, or departments of agriculture in each state.

Quarantine services also keep material in safe areas and do not release it until after it has passed free of pests and disease. For example, during outbreaks of foot and mouth disease, infected farms are quarantined – no material that might carry the disease is allowed to leave the infected area.

Serious penalties apply for breaches of quarantine laws.

Genetic engineering

There is currently much research being undertaken to develop plant varieties that have been genetically engineered to resist pests and diseases. This is a highly controversial technology and there is strong resistance to the use of these plants. At this early stage in their development, genetically modified crops are only suitable for broadacre crops (monoculture). Whether genetic engineering technology fulfils its potential to reduce the use of agricultural chemicals and provide better quality food will depend upon the limitations of the technology, the corporate ethics of biotechnology firms, government regulations and the acceptance of genetically modified crops by both farmers and consumers.

The advantages of genetically modified crops:

1 Resistance to pests or disease
2 Reduced reliance on chemicals
3 Healthier food (for example, a genetically engineered strain of rice contains extra vitamin A. The absence of this vitamin in rice contributes to blindness for many people whose staple food is rice.)
4 Reduced spoilage and longer shelf life

The disadvantages of genetically modified crops:

1 Plant varieties are usually owned by biotechnology companies. In many cases, farmers must pay a licensing fee to grow a particular crop
2 Encourages reliance on limited number of crop varieties (loss of biodiversity)
3 Loss of status as an organic farm
4 Potential for pest resistant genes to be transferred to weed species
5 Unknown side effects of eating genetically modified food

One of the first genetically engineered crops to be widely grown throughout the world was a variety of cotton resistant to the cotton boll worm. It includes genes from the insect toxic bacterium *Bacillus thurengiensis* (Bt). Use of this Bt variety of cotton has resulted in significant reductions in the use of insecticides.

It is interesting to note that the principles of IPM still have to be observed with genetically modified crops grown as a monoculture. At all sites where the Bt cotton is grown, 15% of the area is sown with a non-resistant cotton crop (known as a refuge) that acts as a host area for the beetle pest. If this were not done, the insect would develop resistance to the crop within a few generations, making the expensive Bt cotton strain no more pest-resistant than any other variety.

Pest and disease control in animals

The best way to control pests and diseases in animals is by preventing them becoming a problem in the first place. If a disease outbreak does occur, then swift action is needed to contain and treat the problem. A good farm manager will have an overall picture of the farm system and will not rely totally on chemicals to control pests in his animals. To minimise the chance of pest and disease problems in agricultural systems, there are a number of things that can be done:

Prevent diseases from entering the farm system

When purchasing new stock, look for healthy animals. Where possible, choose disease resistant stock. Newly acquired animals should also be quarantined to minimise the chance of them bringing diseases onto the farm.

Stock need to be inspected regularly and frequently, and sick or diseased animals should be isolated from the general population. Not only will this reduce the chance of the spread of disease, but it will also prevent healthy stock from bullying sick or diseased individuals.

Farm fences should be secure, to reduce the possibility of neighbouring stock entering and introducing disease.

Feed animals an adequate and well balanced diet

The importance of nutrition to wellbeing cannot be overstressed. A well-nourished animal will have an effective natural immune system. Such animals may also be less prone to worm infestation. A badly fed animal is much more susceptible to disease.

It is not enough that stock are given sufficient bulk to satisfy their appetites. The food must supply all needs for fibre, protein, vitamins and minerals. The food must also be of good quality. Poor quality food can, in itself, cause disease (eg fungal spores in mouldy maize cobs).

There is increasing evidence that some foods can protect animals against disease. Garlic, for example, has well researched medicinal properties and is used in some countries as a feed supplement (in powdered form). Another supplement that has proved beneficial is powdered seaweed (supplies many essential minerals and vitamins).

Actively manage grazing and fencing

Stock should be rotated on available grazing. This minimises the buildup of pests/parasites in one area. Rotational grazing also allows feed to renew energy reserves, to rebuild plant vigour and to give long-term maximum production. Good rotational grazing will be timed so that stock arrive at each paddock when the forage has reached the optimum growth stage for consumption.

Stocking rates must be set at a level that can be sustained not only in good conditions, but also in bad times (eg during droughts). Too many agriculturalists become optimistic during good years and stock their paddocks at a rate that cannot be sustained during bad times. Overstocking is simply not sustainable in the long term. Animals from overstocked pastures suffer from poorer nutrition and a higher incidence of health problems. Pastures should also be checked regularly for poisonous plants or those likely to cause internal obstructions to animals.

Fences should be well maintained not only to contain stock but also to reduce the likelihood of animals injuring themselves on sharp wire, rusty posts, etc. Animals should also be fenced out of dangerous areas. For example, large animals can get stuck in boggy areas. These areas are also a breeding ground for a number of insects and diseases (eg liver fluke, mosquitoes, heart worm). Such areas should either be drained or fenced off. Animals should also be fenced out of steep or unstable banks where there is a high risk of injury.

Control insects

Insects such as flies and midges spread many diseases. These insects are attracted to the farm by the warmth of the animals and the smell of manure. Where animals are penned, regular removal of the manure can reduce the insect population. Manure should be stored well away from animal pens and removed regularly in order to minimise the threat from insects. Some insects lay their eggs in manure.

Still water can also attract insects. Avoid leaving buckets or other containers around which can collect rainwater. A slow burning, smoking fire will deter many insects (as a short-term measure only!).

Dip animals

Dipping prevents parasites from biting their hosts. Dipping should be done regularly and with the correct mixture to be effective. For agriculture to be truly sustainable, it is advisable to minimise the use of artificial and long-lasting chemicals. Where parasites are a problem in certain animals, diatomaceous earth has been used, in the belief that it reduces the need for insecticides. Proponents of its use claim that it acts as a physical irritant to insect larvae. It can be applied externally or added to feed rations (provided it is used correctly). The particle size of the product is important – if it is too coarse, it is not effective. Diatomaceous earth has a well-documented ability to reduce insect problems in stored grain, but its use in animals is still unconventional and more research is needed to substantiate its effectiveness as a treatment in animals.

Figure 5.4 Diatomaceous earth is used to control insect pests in stored grain.

Vaccinate

Vaccination involves injecting a small amount of the disease organism into the animal which, in turn, will produce antibodies to overcome the disease. The animal is then protected from future infections by these antibodies.

Avoid stressing stock

Stress is any stimulation that puts strain on an animal's body. A stressed animal is much more susceptible to disease. Pigs, for example, are highly stressed if exposed to very hot or cold conditions. Animals grown in an artificial environment may produce more, but may also be more susceptible to disease. Routine tasks like dipping might also cause stress, but this needs to be balanced against the benefits.

The farmer should generally try to keep animals in the most natural environment possible. This is not always easy as some types of livestock just don't produce their full potential under 'free range' conditions. Animals should not be exposed to extremes of temperature or other environmental conditions. Handling facilities should be safe and efficient not only for the handlers, but also for the animals. Staff need to be properly trained in how to handle stock without causing undue stress. The farmer needs to balance stresses caused by nature with stresses caused by an artificial environment and then develop an appropriate environment for the livestock.

Practise good hygiene

Good hygiene is an important part of disease prevention. Manure and droppings need to be separated from animals in intensive operations. Animal bedding should be clean and food and water supplies should be free from contamination.

6

Sustainable natural weed control and cultivation

What is a weed?

The generally accepted definition of a weed is any plant that, for some reason or other, is unwanted in a particular position. Any type of plant has the potential to be a weed. Common reasons why people do not want species to grow include:

- Competition – weeds can compete with your desired plants for space, light, nutrients and moisture
- Safety – some plants may be poisonous or cause allergies (eg St John's wort, bathurst burr, parthenium weed), others may have spines or spikes (eg giant devil's fig, currant bush and Chinese apple) or sharp grass seeds that can injure animals and humans (eg spear grass)
- Harbouring or hosting pests and diseases – some plants may act as hosts or as attractants to pests or diseases, while others may provide a safe haven for pests such as rabbits and foxes
- Tainting – some weeds (eg capeweed, wild garlic) can taint the taste of meat and milk from animals such as cattle, sheep, goats and pigs
- Contamination – plant parts, particularly seeds, can get caught up in clothing, or can contaminate produce, such as grains, or get entangled in animal fur or fleece, or fibre crops (eg cotton), or in hay
- Interfere with cultivation – some plants can become entangled in machinery, making tasks, such as cultivation, mowing, or machine harvesting difficult, and possibly damaging machinery
- Soil erosion – some weeds are very competitive and will easily shade out other more desirable plants; if the weed is only seasonal (eg an annual, or dies back during winter), it may leave exposed soil that may be easily eroded
- Aesthetics – some plants may be weeds simply because they look bad, or they have offensive odours

- Environmental – these are plants that invade native vegetation, displacing the indigenous species; they can severely affect local flora and fauna populations (eg lantana, camphor laurel, Singapore daisy and water hyacinth).

Controlling weeds

Once it is determined that a particular plant or group of plants is a weed, we need to select a suitable method to control it. Chemical methods are the mainstay of weed control on most farms. Weedicides certainly give a quick result, but also have the following problems:

- They can be quite expensive
- There may be legal requirements with regard to their use and storage, and training of operators
- Chemicals can damage other plants, especially if they are applied in windy or hot conditions
- They can wash off in rain and either don't work, or they may run into other areas, causing damage to other plants
- If you get the concentration wrong, chemicals can actually promote rather than deter growth; some blackberry killers, for example, used at low concentrations cause more rapid growth
- High concentrations can poison the ground and, in extreme cases, prevent further plant growth
- Chemicals can be harmful to animal life including humans, domestic pets, birds, fish, and soil life
- The manufacturing processes involved in making the weedicides can cause pollution problems
- Old weedicide containers pose a safety risk unless carefully disposed of

Even in 'sustainable farming' methods such as conservation tillage, chemical use is common. It would be extremely difficult for most farmers at present to completely stop using such chemicals; however there are some non-chemical control methods that can be readily applied and which can significantly reduce the farmer's dependence on weedicides. You should first:

- Know the weed or weeds you are dealing with
- Know how those varieties grow, and what conditions they do and don't tolerate
- Then create conditions which they don't like

You need to consider whether you want to kill or just control the weeds. When you know these things you can consider which method is best for your situation.

Ways to control weeds without chemicals

Check soil condition

Weeds are much more of a problem if your soil is infertile, poorly structured, or is regularly disturbed in some manner, such as by excessive cultivation. Keep your soil in good condition (eg fertile, properly drained and friable) and your plants will compete strongly with weeds.

The first step to control weeds is to improve your soil:

- Use soil ameliorants, such as lime or gypsum where necessary
- Add organic matter regularly (as a mulch or dug in before planting)
- Fertilise your plants regularly
- Improve drainage if necessary

A good cover of plants will also reduce the likelihood of weed invasion. Bare, disturbed soil creates the ideal opportunity for weeds to become established.

Minimise sources of weed seeds

The next step is to control the source of weed seeds. If you stop weed seeds getting onto your property, you will stop most weeds from becoming established.

- Give first priority to removing any weeds in flower before seed is produced. It may be necessary to look outside of your property to see where the weed seeds are coming from. They may be blown or carried in from elsewhere. You may possibly be able to control this by such means as slashing down weed plants immediately adjacent to your property, perhaps on road verges or adjacent bushland verges.
- Be careful not to bring any contaminated plants, soil or mulch (or anything else containing weed plants or seeds) onto your property. Materials, such as fresh manures or grass hay to be used as mulch, should be composted before being spread, to kill off as many as possible of the large numbers of seeds that they usually contain.
- When bringing in animal stock from elsewhere, place them in a confined area for a week or two, so that any weed seeds they may have in their digestive systems are passed out, and can be easily controlled as they germinate, rather than being spread over a much larger area, making control more difficult. This has the added advantage of keeping the new animals initially isolated from your existing stock, in case they have pest or disease problems which haven't been noticed. If you are buying fleeced animals such as goats and sheep from areas where problem weeds are present, these can be obtained shorn, so that the possibility of transferring weed seeds in the animal's fleece is greatly reduced.
- Any machinery that has been used in areas with weed problems should be hosed down before being moved to a weed-free area.

Cultivation

Cultivation (ie digging or turning the soil with a spade, hoe or engine-driven cultivator) will disturb weed growth and, in the case of annual weeds, often kill the weed.

- Young weeds are damaged more by cultivation than established weeds
- Do not water the soil after cultivation (the hot sun kills exposed roots)
- Some weeds will die quickly when you cut the top from the roots (others will regrow from the smallest piece of stem or root lying in the soil)
- There will always be some hard to kill weeds which need removing by other methods, some weeds are almost impossible to control, even by hand, unless you have a lot of patience
- Some weed seeds germinate very quickly after soil disturbance

Mulching

A popular weed control method is to suffocate the weed (block out light) and/or put a physical barrier over it which it can't grow through. This is most commonly known as mulching. What mulching does is to kill weeds simply by smothering them. The weeds are deprived of light and in order for them to grow they have to break through the barrier formed by the mulch. A mulch can take the form of almost anything, but the more popular ones are:

- Wood shavings and chips
- Pine bark
- Hay or straw
- Grass clippings
- Leaves
- Newspaper
- Carpet underfelt
- Cardboard
- Seaweed
- Sawdust

The depth of the mulch will be determined by the weeds that you are trying to control. Vigorous weeds will need a greater depth of mulch than perhaps small annual weeds. Most weed seedlings will require a depth of mulch of 8–10 cm over the top of them. This thickness can be reduced by mowing the weeds then covering with a thick layer of newspaper (perhaps 15 to 30 sheets thick) before laying down the mulch. Grass clippings or hay that may contain large numbers of seeds should be avoided.

Mulch mats – these are also known as weed mats. They are usually made of a closely woven fabric perforated with holes large enough to allow water to penetrate, but small enough to prevent most weeds from growing through.

Biological weed control

This involves introducing natural predators into an area to attack weeds. It is a method which has been used occasionally with dramatic results, but which can backfire if the full implications of introducing something new into an environment are not understood.

Figure 6.1 Prickly pear (*Opuntia* sp.) – this weed was a massive problem in Australian pastures until the moth *Cactoblastis cactorum* was introduced as a biological control agent. Prickly pear is now rarely seen.

Examples of biological control of weeds:

- Prickly pear (*Opuntia* sp.) – This cactus was a severe problem in the past in New South Wales and Queensland, but was brought under rapid control in Australia by introducing a parasitic moth (*Cactoblastis cactorum*) which has a grub that attacks the plant.
- Blackberries – A rust (fungal) disease was introduced into Australia in the 1980s in an attempt to control blackberry weeds. Though this has had some effect, to this stage it has only been a mild deterrent.
- Water hyacinth – Insects have been used to control the spread of water hyacinth in the United States and Australia.

Grazing

Regular grazing of larger areas will effectively control many weeds and keep grass down to an acceptable level. Grazing animals should be well fenced in to ensure that they only eat what they are supposed to. Simple electric fences can create a temporary enclosure in areas that are not regularly grazed. Smaller grazing animals (even poultry will graze out many weeds) should be protected from predators (eg dogs, cats, foxes). An adequate supply of clean water is also vital. The manures of such animals can supply valuable plant nutrients to your property.

Goats

Goats are excellent for controlling weeds. They eat virtually anything. The trick is to keep them in the area where the weeds are and keep them away from any valuable plants. Here are a few hints if you're considering a goat:

- For small properties it is better to borrow a goat than buy one. Otherwise, when you run out of weeds, feeding it can become a problem.
- Goats are best used to keep a wild area under control on a large property or to clean up an area prior to growing a crop.
- Goats are very strong, they can break small gauge chains, eat through ropes and pull stakes out of the ground. You will need to have them strongly fenced or use a heavy chain and tie them up to something very solid such as a fence post or large tree.
- Goats will stand on their back legs to reach plants; they will eat all types of plants and even strip the bark off trees.

Other grazing animals

Sheep can also be used for grazing, but can be a little more choosy in what they are prepared to eat. Poultry will also eat a variety of weeds, and cultivate the soil by scratching. Penned pigs will also cultivate the soil with their digging. Wire netting is sometimes placed on the ground in a poultry run to stop hens digging up the soil too much.

Chemical control of weeds

Prior to the development of purified toxic chemicals at the beginning of the 20th century, sea salt was the only widely available herbicide (weed killer). During the 1940s selective herbicides were developed and today herbicides come in a range of products of varying toxicity. Since the 1970s, the level of herbicide tolerance and resistance has steadily increased, due to continual use of chemically similar herbicides. In recent times, it has also become obvious that herbicide reliance requires ongoing use without ever reducing the problem. Herbicides should be used in conjunction with other weed control methods.

Herbicides

Herbicides can present a significant danger to non-target organisms. In particular, residual and foliar applications have the potential to kill desired plants, poison animals, reduce soil micro-organisms, evaporate into the atmosphere and enter the watertable.

Herbicide use is controlled by governments through laws and regulations. Before using any herbicide, check with your local authorities that it is authorised for use in your area. Always read the label before use and always follow the application rates and safety instructions as directed.

Herbicides can be liquid or granular preparations applied to foliage, stems or the soil. Their action can be systemic or contact and they can be residual or non-residual in the environment. Most herbicides are non-selective and kill most plants they contact.

Sea salt

Easily the cheapest of all herbicides, it is sold as non-iodised or cooking salt and is effective against many leafy weeds. Dissolved in water, it is taken up by the plant systemically. Problems with residual salt in the soil prevent its widespread use.

Glyphosate

Sold under trade names such as 'Zero' and 'Roundup', glyphosate is a systemic, non-residual and non-selective herbicide spray that is widely used around the world. Its systemic action means that it will not eradicate many plants with bulbs or other food storage mechanisms. It acts by interfering with enzyme activity in plants. Because there is no equivalent enzyme activity in animals, it is considered 'safe' for use. In recent times, some weeds have begun to develop resistance to glyphosate.

Contact herbicides

Contact herbicides include the non-selective, non-residual compounds paraquat and diquat. These herbicides must be applied evenly to the target plant to be fully effective. They are poisonous if inhaled or swallowed.

Woody weed herbicides

For persistent weeds such as blackberries and trees, it may be necessary to use stronger chemicals. These chemicals are non-selective and systemic. Examples include amitrole (Weedazole®) and triclopyr (Garlon®). In some areas, the use of these chemicals is now restricted to licensed operators.

Residual herbicides

These systemic herbicides are applied to the soil to prevent weed seeds germinating, usually during cultivation or irrigation. They include amitrole and simazine. They have long-term persistence in the soil and are likely to enter the watertable after periods of heavy rain.

Selective herbicides

These chemicals are systemic and only poison one type of plant, either broad-leafed weeds or grassy weeds. This means they can be sprayed on crops with unsuitable foliage. Some of these chemicals, including MCPA and 2,4 D, act by interfering with hormone activity in plants (and animals). They are highly toxic and residual in the environment. In a number of countries, many of these chemicals are now either banned or strictly controlled.

Herbicide additives

Dyes

Non-toxic coloured dyes such as food dyes are sometimes added to herbicide sprays. This reduces wastage by allowing the operator to see where they have already sprayed. It also acts as a warning to visitors that herbicide has been used in the area.

Surfactants

These are additives that help keep the herbicide mixture on the surface of the plant leaf. They include soaps (beware – soap suds can block spraying equipment) and various commercial preparations.

Other weed control methods

There are other non-chemical methods of controlling weeds, however most tend to be 'drastic' techniques which not only kill weeds, but can kill everything else at the same time. Such techniques might be useful in some situations such as clearing a new area prior to planting, or weed control on fence lines. They are often unsuitable for treating existing crops.

Mowing / slashing

This involves regularly cutting the tops from the weeds. The cut foliage should be left to rot and return nutrients back into the ground. If the weeds are tall when cut, the foliage will act as a mulch, slowing regrowth of weeds. Cutting close to the ground does more damage to the weeds than cutting high. Whipper snippers (ie brush cutting machines) are ideal for this.

Flooding

Flooding an area will kill a wide range of weeds (but not all). This is sometimes used on flat sites prior to planting.

Solarisation

Large sheets of clear plastic are spread over the surface of the ground in warm weather. Heat generated under the plastic can be great enough to kill many types of weeds. The plastic can then be removed (perhaps after a couple of weeks) and the area planted. This technique will also often kill other pest and disease organisms, but it is only suitable for relatively small areas.

Burning

Flame throwers are used by some government bodies and farmers for killing weeds on boundary fences, or large clumps of weeds such as blackberries. Be sure not to burn the fence. Care should be taken with this technique as the heat generated can be quite considerable and can damage other plants, as well as causing nasty burns to the operator.

Controlled burning of areas is sometimes used to suppress weed growth. Fire should only be used by skilled operators, and never in an area where there is danger of a bushfire occurring. It is very important that you check with your local fire brigade for advice and to ensure that burning-off restrictions are not in force.

Avoid creating fire piles or bonfires in areas that are to be used later for growing crops. The intense heat of a fire maintained for any length of time in one position can sometimes scald the earth surface or kill beneficial soil organisms, making it difficult to grow any plants in that area for some time. As well, the concentration of particular plant nutrients such as potash (potassium) can also have a major effect on plant growth.

Changing soil pH

Every plant has its preferred pH range. Changing soil pH can sometimes be used to control some types of weeds. For example, by raising soil pH you can discourage growth of sorrel (*Rumex* sp.). Adding organic matter to the soil will also gradually cause sorrel growth to slow down.

Harvesting your weeds

Instead of simply getting rid of them, there are a few weeds that can be harvested for your own use. One of the most common is the dandelion (*Taraxacum officinale*). The leaves can be used as a salad green, the flowers in pot pourri, and the roots roasted and ground as a coffee substitute.

Weed control with hot water

This environmentally friendly method involves spraying extremely hot water onto weeds. Impressive results have been obtained through research at Florida Citrus Research Institute and California State University (Fresno). In most tested cases, both broad-leafed weeds and grasses have been effectively controlled. The equipment for this technique is available from 'Spray Tech' at Nerang, Queensland, Australia.

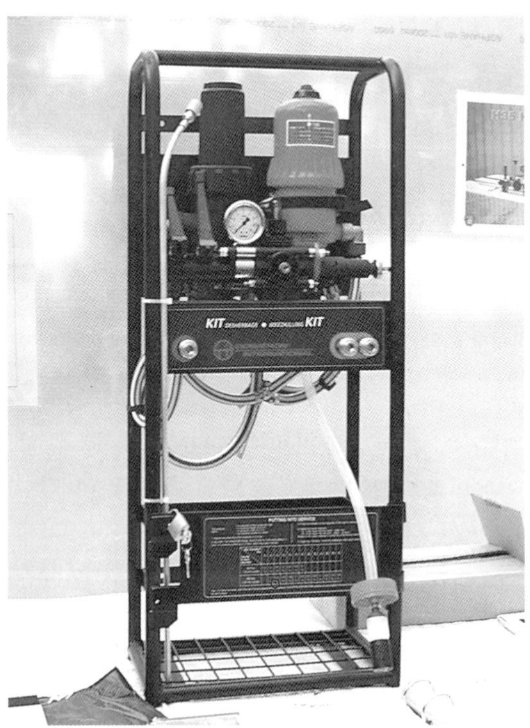

Figure 6.2 This weed control steamer is used to kill weeds without the need for conventional weedicides.

Plants that take over

Many garden plants can become weeds. It is wise to think carefully about the vigour of plants you use in your farm garden and, if you want to avoid creating weed problems elsewhere on your property, avoid planting the following types of plants:

- Plants that go to seed freely; during the growing season, some plants set prolific amounts of seed, which mature and fall to the ground. These seeds may well germinate during the growing season, or will germinate later on. Whatever the time the seeds take to germinate, there will be a great deal of them and you can be faced with a mass of new and unwanted seedlings. This can be a real problem if you have crops growing nearby.
- Vigorous creepers that take root (eg periwinkle, ivy); this type of plant readily spreads across shrub borders with long, branching shoots. These will tend to wind themselves around the other plants and will also readily root into the ground if they come into contact with it. This means that the plant is able to spread very rapidly. Once these roots have become established, it can become difficult to control the plant's growth, and the only sure way is to dig the plants out.
- Plants that spread by suckers or underground roots (eg raspberry, willow, bamboo, couch grass, kikuyu, many perennials). One of the main features with this type of plant is that you do not know about the problem until it appears. As the plant is spreading underground, the roots can travel some distance before they surface. Just cutting the growth off where it appears will not solve the problem as the roots may well reshoot from the buried root. You need to dig the root or sucker up along its total length.
- Plants that have very rapid seedling growth; even a few seeds may grow very rapidly and soon begin to bear seeds themselves.

Environmental weeds

Many garden and some crop plants can readily escape into bushland areas, where they can compete or even completely take over from the native vegetation. The stability or balance of natural systems can be upset, causing radical changes and, as a result, habitats for native fauna can be severely damaged. Environmental weeds can greatly reduce the variety of species present, and also may reduce access and recreational use by creating impenetrable barriers of twining plants or dense thickets.

If suitable pollinators and seed-dispersing animals are present, exotic plants will flourish without the pests and diseases that kept them in check in their original country, all at the expense of the native plants. A European broom, for example, could replace a native wattle. Plants may spread by being dumped (a common problem along railway lines and in bushland adjacent to residential areas) or by seed, often carried by birds. Another problem is that introduced plants can sometimes cross-pollinate with the local native (indigenous) plants, for example Grevillea species and cultivars. This interbreeding results in hybrids which interfere with the natural evolution of the indigenous plants.

In the middle of suburbia, growing these types of plants is not generally a great problem, but on farms, particularly if there are areas of native bush nearby, there is a strong chance of plants escaping. Many plants commonly used in windbreaks (eg tree lucerne) can also pose similar problems. Many road and rail reserves, foreshores, bushland areas, and national parks are now infested with environmental weeds. The situation can be so bad that all weed control methods may be needed – including hand, chemical and biological control. The loss of natural bushland is a great cost to the community, in lost educational and recreational opportunities and the loss of indigenous plants and animals, as well as the public funds used (or not used) to control the problem.

Some plants to avoid

Few plants become problems in areas that have climatic extremes such as snow, desert, etc. Plants tend to become troublesome when they are introduced into areas that have a similar climate to their native habitat. The following list, while not comprehensive, includes some common plants that have become environmental weeds in some areas of Australia. The table serves to highlight the importance of caution when selecting any plants for culture. It is important to remember that a particular plant may be a weed in one area but not in another. If you are not sure about which plants are problems in your area then contact your local council, or state department of conservation, or perhaps a local conservation society or tree planting group.

Table 18 Plants to avoid

Plant name	Comments
Acacia baileyana (Cootamundra wattle)	Native of south-east NSW that has become extensively naturalised in other parts of Australia; widely grown in gardens, and as a street tree
Acacia longifolia (sallow wattle)	Heavy seeder; germinates readily and doesn't need fire; native that has naturalised extensively outside it's natural range
Albizzia lophantha (Cape Leeuwin wattle) (syn *Paraserianthes lopantha*)	A rampant seeder, a problem, particularly in sandy and coastal areas, fire enhances germination of the seeds
Chamaecytisus proliferus (tree lucerne, tagastaste)	A popular fodder tree that can also become invasive in some native vegetation types
Chrysanthemoides monilifera (boneseed)	Rampant seeder, with seed remaining viable for some time; a problem, particularly in coastal areas; easily pulled out by hand, but follow-up checks are necessary for several years to destroy new seedlings
Cinnamomum camphora (camphor laurel)	This is often a prized feature tree in temperate gardens, the fragrant timber is valuable for wood work; in Queensland however, it has developed into a vigorous and serious weed
Cotoneaster sp.	Produces berries that are spread rapidly by animals and birds
Cytisus scoparius (English broom)	Heavy seeder that rapidly spreads in disturbed areas, common on some road verges

Table 18 Plants to avoid (continued)

Plant name	Comments
Hakea suaveolens (Sweet hakea)	A West Australian native that has become extensively naturalised in the eastern states and some overseas countries; it rapidly spreads, particularly in coastal areas, with prolific germination after fires; often planted in windbreaks
Hedera helix (English ivy)	Invasive creeper/climber with aerial roots; a major problem in moist shaded spots, particularly in moist native forests, where it smothers other vegetation, including trees
Ipomoea indica	Blue Morning Glory can be a problem in many parts of Australia. Other morning glory species (*Ipomoea* spp.) can also naturalise and become weeds in some areas, despite their popularity as cultivated garden plants
Lantana camara	An attractive, easily managed shrub in temperate areas, a rampant weed in tropical regions
Lavendula stoechus (Italian lavender)	Grown widely as a hardy herb plant but declared a noxious weed in South Australia and Victoria
Lonicera japonica (honeysuckle)	A rampant scrambler/climber that smothers other vegetation
Pinus radiata (Monterey pine)	Large tree that displaces native vegetation, particularly in open or disturbed woodland areas
Pittosporum undulatum	A native small tree that has spread extensively out of its natural area of occurrence, displacing many indigenous natives; the seed is spread easily by birds
Polygala myrtifolia	Spread widely in sandy places, such as beach areas, particularly in Victoria, SA and WA
Prunus cerasifera (cherry plum)	Readily spread by animals, particularly birds and humans; common on some roadside verges, and along streams
Salix spp. (willows)	A major problem in moist areas; they can completely replace indigenous vegetation along streams, choking the streams in the process; *S. cinerea* will spread by seed, the others spread easily by vegetative means
Salvinia molesta	A water weed that can choke lakes and streams, a problem in warmer areas (eg NSW, Qld)
Tradescantia fluminensis	A rampant groundcover/creeper in moist, shaded positions; it self layers readily; similar plants which can also create problems include *Commelina cyanea* and *Tradescantia albiflora*
Thunbergia alata (black-eyed susan)	Popular climber in temperate Australia, but a serious weed in Brisbane and further north; other species of the genus 'Thunbergia' are also potential weeds, and should be used with caution
Verbena spp.	Grown as a garden plant in temperate climate, verbenas can become a serious weed in subtropical areas; a rampant creeper, that suckers freely
Vinca major (periwinkle)	A problem in moist, semi-shaded areas

Noxious weeds

In all states of Australia and in many other parts of the world, there are laws covering the control of declared noxious weeds. Noxious weeds are those that present a serious threat to agriculture, the natural environment and/or waterways. All landholders should be aware of the noxious weeds in their area. After weeds are declared noxious by the local department of agriculture, it is the responsibility of landholders to eradicate any noxious weeds on their property. Failure to do so can result in serious penalties.

7

Management

A farm is like a machine, fuelled by various resources, such as pasture, manpower, soil, water, equipment, crop plants and animals. As the resources become depleted or are used improperly, the profitability of the 'machine' decreases, wasting resources and diminishing in both production and financial terms.

Sustainable agriculture is a long-term answer to this scenario. As with any business enterprise, long-term goals should be in place to give the business a direction that will see it running successfully for generations to come. What often appears as short-term loss, decrease in production, or expensive investments may be the very things necessary to keep the business viable into the future. Such is the case with sustainable farming.

If a method of farming is not profitable, it is not sustainable. Sustainable farmers recognise that whilst their returns may be lower than those of conventional farmers, their costs are also lower due to lower inputs of expensive items such as chemical fertilisers and pesticides.

Changing an existing farm to a sustainable property

It can be both disruptive and expensive to change an existing farm from a 'traditional' operation, to a 'sustainable' one. The change should not, however, be expected to take place overnight. A realistic approach is often to convert to a low input-sustainable system over a period of 5–10 years, one paddock at a time. There is no sense in trying to make a farm environmentally sustainable overnight if in doing so you lose both financial sustainability and ownership of the farm.

Making the change will always involve some trade-offs. For example, the quality or quantity of production might decrease but the cost of production might also be reduced. In balance, the profitability remains the same. The farm will have lost very little in the short term in order to achieve sustainability in the long term.

Other changes that can be gradually introduced might include:
- reducing or eliminating the use of pesticides (eg by embracing integrated pest management or biological controls)
- changing cultivation methods to reduce damage to soils
- changing or rotating the varieties of plants being grown as crops or pastures
- changing the types of animals being farmed, or managing animals in a different way (eg restricting their movement)

Perhaps the most important decision a farmer will make is deciding what to grow. Traditional products such as wheat, sheep and cattle have long been considered 'safe' options in terms of having an assured marketplace but more and more traditional farmers are facing bankruptcy. Farms that produce one product are not protected against changes in market trends. Diversification allows the farm to rely on more than one product, thus supporting it through times of market fluctuation. By diversifying and producing several different products, the disadvantages of a monoculture and all the associated problems that arise from weather, difficult growing seasons and market changes are minimised.

Diversification advantages include:
- providing a buffer against financial catastrophe if one of the products fails
- averaging profits and losses over the different products
- providing an opportunity for value adding by combining two products (eg cheese and herbs)

Diversification will allow the farm to be sustainable, both ecologically and financially. While financial considerations will always influence the final product mix, all farm products must be assessed in terms of their effect on the environment. Some things that should be considered when deciding what to produce include:

- choosing plants and animals that are more drought tolerant, require less maintenance and feeding, attract fewer pests, are disease resistant, and have less physical impact on soil and resources
- the use and/or creation of alternative food sources, eg fodder tree plantations, roadside grazing
- dual-use products (eg plants that provide fruit can also provide fodder or support bees; sheep that produce both wool and milk)

There are socioeconomic and political considerations when converting to sustainable farming as well. For example, many government programs still favour monocultures over mixed cropping. Many financial institutions feel that agricultural chemicals are the best tools for protecting money they have loaned out and discourage other methods of pest management.

The key to successfully converting from a conventional to a sustainable farming system is good planning. Change can be a destructive factor in an agricultural system but planning ahead means change is not a surprise to the manager.

The Rodale Institute conversion experiment

The Rodale Institute is a long established centre, which aims to achieve a regenerative food system that renews environmental and human health. The centre's philosophy is that healthy soil means healthy food, which means healthy people.

In 1981, The Rodale Institute commenced conversion of a conventional cropping system to a more sustainable system. Prior to conversion, chemical fertilisers and pesticides were used on the property. These were removed from the farm system. In a separate system, animal manure and ploughed down legumes were used to supply nutrients. Crop rotations were initiated. The experiment found:

- When initiating crop rotations in soil that has been farmed conventionally, it is better to commence with a leguminous hay or green manure crop. When grown with a small grain nurse crop, these plants will suppress weeds and improve soil fertility/structure for subsequent crops.
- The addition of organic matter to the soil was found to have a marked effect in a reasonably short period of time. Soil with more organic matter generally has better structure, more topsoil (due to reduced erosion), better nutrient availability and more soil life.
- When synthetic fertilisers and pesticides were removed from a conventional cropping system, there is a transition period where a new plant-soil ecosystem becomes established. For example, corn yields in the newly established sustainable system did not equal those of conventional systems until the fifth year of the experiment. From that time onwards, however, the sustainable system has performed equally well or better. The lower yields in the transition period appeared to be due to a lack of available soil nitrogen and increased weed competition.

The Rodale Institute has also found that the amount of weeds farmers will tolerate in a crop field can influence their success in converting to a more sustainable system. For example, up to 500 kg per ha weeds can be allowed before the yield of a corn field is reduced.

New farm products

From time to time someone discovers and promotes a 'fantastic' new crop or animal to be farmed. These include animals not previously farmed, such as deer and emu, or produce popular in another country, such as exotic fruits and vegetables. In most cases these are things that do have a lot of potential, but along with the chance of high profits, there also comes a risk of big losses. If a new product must adapt to a new climate, there are usually problems associated with the fact that too many questions are unanswered, leaving the producers to learn by trial and error. This is where the risk lies, but it can be reduced if the products are well researched. When assessing a new farm product, consider:

- What will the demand for this product actually be? Once the initial novelty dies down, will the market continue?
- New products do not have the same expertise or experience to support their production that established products have. Is your farm sufficiently diversified to support the trial and error period necessary to establish a new product?

- How does the product fit in with the long-term sustainability of your enterprise?
- Because it is scarce, the cost of purchasing stock may initially be high.
- Markets are not established, so more work and expense will be involved to sell the product.

There is a history of failed farms with many of what were once new products. Some of these may have now found their place in mainstream agriculture, but during the transition from something new to a respected industry sector, many people may have been badly hurt. For example, many hydroponic farms failed miserably in the 1970s in Australia; today some of the country's most profitable and sustainable agricultural enterprises are hydroponic ventures.

Pre-planning

The first step to establishing any farm enterprise is planning. Planning starts with research, both into the property in question and the products to be grown. Gather together factual information and undertake some research into the property that is to be used. Even if you have been farming the property for generations, an assessment should be made.

Planning involves monitoring relevant factors, and correcting negative trends before they have a chance to seriously impact upon a farm's sustainability. Sustainable farming can be planned on a macro or micro scale. A macro scale is a cooperative approach to sustainable farming across several adjoining properties. While the properties may have different products, they can work together to make things such as windbreaks, nature corridors or water retention areas that benefit more than one property. On the micro scale, one farm, or even one part of a farm can be planned for.

To avoid disruption and possible economic problems, plan before implementing any changes. The changes may involve redesigning the farm, incorporating major and minor changes, both immediately and in the long term. Develop a master plan then gradually work towards it. But remember that a master plan should, like the farm itself, be a living thing. Adapt the plan to changes that time and experience will make apparent. The plan should be referred to periodically to check progress, note changes and keep the business goals in view.

Considerations

Planning can often be the most difficult stage of any development. Transferring a bright idea into a workable plan can be frustrating. To this end, there are some general considerations that can help to get the ball rolling.

Products

- Grow crops or animals which are appropriate for soil, water and climatic conditions
- Grow products which are properly serviced (eg adequate storage, transport facilities, production machinery)
- Have financial, manpower and other necessary resources
- Integrate crop and livestock nutritional needs

- Ensure you have an accessible market
- Integrate the effect of new crop/livestock (income and expenditure) on total farm cash flow

These points should be considered in tandem with all of the following considerations.

Farm activities

Look at the alternatives available when dealing with:

- control of pests and diseases (including vermin)
- control of weeds
- rotation of crop/pasture use
- contingency plans for drought, flood and other catastrophes
- land use capability (eg be conservative with stocking rates)
- condition of physical resources (equipment, materials, land, etc)
- conservation of physical resources (eg don't waste water)
- review of production systems (eg improve/develop better ways of production)

Off-property effects (surrounding properties or communities)

- Avoid land use conflicts
- Warn neighbours of anything that might affect them (eg burning off)
- Care for the natural environment – conserve flora and fauna, protect remnants of natural vegetation, control domestic pets, protect fish habitats, develop wildlife habitats, restrict livestock access to natural watercourses, etc

Information

- What new skills need to be developed (eg training courses on production techniques, attending information sessions)?
- What new information resources will need to be accessed (eg subscribe to publications, use consultants, joining organisations)?
- What practices will need to be monitored and/or reviewed? Routine testing or inspections may need to be implemented to assess practicality, possible improvements and sustainability.

Biological and climactic considerations

- The biology (soil life, fertility, pests, diseases, pastures etc) of a property can change over time; regular control, introduction and manipulation of biological organisms need to be adjusted to current conditions
- Plan to minimise chances of pest, disease or weed problems; introduce measures that are sustainable and compatible to the products to be harvested
- Continually improve soil fertility and ensure appropriate soil life
- Note seasonal changes to the property in terms of weather patterns, wet and dry spots, existing and possible microclimates to be developed; try to identify the strengths that can be exploited and the weaknesses that need to be upgraded; estimate the cost of any changes and the savings of changing the use of an area (ie less irrigation, lower maintenance, etc)

Management

- Adapt and change to more appropriate production systems (eg tillage, harvest, storage, pest control)
- Make the farm layout as efficient and useful as possible (eg existing and incorporated windbreaks, seasonal and permanent water sources)
- Adjust record keeping systems to monitor areas of importance for both financial and ecological sustainability (eg record problems when they occur, to be investigated when time allows; record breakthroughs in previous problem areas)

Socioeconomic options

- Calculate the costs and savings that can be expected in the shift to the sustainable farm. Sustainable farms often use less fuel, fertiliser, chemicals, etc. and hence have lower production costs.
- Plan to get more involved with the community. Find out what programs and organisations exist that may be of help. Look toward developing a more cooperative approach to property use and management. Don't try to reinvent the wheel, use the information and experience that others can provide.
- Sustainable farms may be more viable on a small scale than 'traditional modern farms'. Estimate what land area you will actually need to be sustainable and consider leasing or share cropping extra land available. This type of action is supportive of the local farming community and will help to maintain rural population numbers. The result is better infrastructure and more services available.

Political/legal considerations

- Legislation sometimes restricts the way in which something can be farmed. Ensure you are aware of any legal ramifications in terms of quarantine, chemical treatments, interstate transfer laws, etc.
- Ensure you are aware of local zoning regulations; especially in areas close to the rural/urban interface, some restrictions can be prohibitive to what and how you produce.

Monitoring and reviewing the farm system

Sustainable farming requires continuous monitoring of the condition of the farm. Continually check for the following, and adjust your management and long-term plans whenever a problem is identified.

Deterioration of soil

- Lowering of organic content
- Lowering of EC (electroconductivity)
- Changes in pH
- Preliminary signs of erosion
- Salinity (white caking on soil surface or indicator plants)
- Reduced plant growth

Deterioration in water quality

- Increased EC
- Algal blooms
- Clarity or colour changes

Weeds and pests

- Watch for dramatic changes in their populations
- Monitor for changes in susceptibility of pests or diseases to treatments (this can indicate resistance is developing in new generations of the pest)

Crop and livestock health

- Watch for deterioration in crop or livestock health, including increased susceptibility to disease; discolouration of plant foliage, etc
- Be aware of any drop in crop or livestock yields

Socioeconomic considerations

Sustainable agriculture can be seen in many different lights. Some managers choose to pursue sustainability to the full extreme in enterprises such as permaculture systems, where the human inhabitants become an extension of the system. Some see sustainable agriculture as a shrewd business decision, to ensure the longevity of the business enterprise. Regardless of personal views, decisions must be made in regard to both social and economic effects of the system.

Profitability

Liquidity needs to be maintained within the enterprise. Will the production mix and levels support the ongoing operation of the farm? Can what has been identified as ecologically sustainable support the property financially? Will the changes to the property force a severe change in lifestyle on the property owner in terms of income? These are all important questions that need to be answered.

Measuring the short-term costs against the long-term gains will often show that the financial cost of sustainable farming in the short term will return both financial and environmental benefits in the long term. When considering changes to the property, ask the following:

- What is the initial cost of the change?
- What will be the loss in production/potential income?
- What will be the savings in labour, maintenance, materials?
- How long will it be before the new product is in full production?
- How will this change affect the ability to service existing debts?
- Will additional debt be required to make the desired changes?

Sustainability will always come down to weighing the short-term versus long-term economic costs. A good example is the inclusion of stands of native forest on a property. Reserving areas for specific types of native timbers can have both short and long-term financial benefits. Located properly, a stand of timber can provide a windbreak, reduce soil erosion and increase water retention rates. While not providing a monetary return, these things will affect the cost of production by reducing wind damage and irrigation costs.

In the longer term, the trees can be seen as an investment for retirement or for family inheritance. As native timber sources become more scarce, the value of harvested timber will increase. While the length of time of the investment is very long term (80–100 years or more), the return on the investment would likely be superior to that of things such as stocks or property. Planted over the years, this resouce could be seen as continual.

Aside from long-term investments, moving toward sustainability also has short-term economic benefits. Business investments create tax deductions, so tax savings on any money spent should be factored in. Sustainability also goes hand in hand with diversity and, as mentioned before, provides protection against poor markets and falling prices in a specific area of production, should this occur.

Social aspects

Business is all about making advances and making money. However, with all businesses the time of making money at all costs is gone. Being a 'responsible corporate citizen' applies as directly to farming as to any other enterprise. This is why the move to sustainable farming is so important.

Arable land is fast becoming a scarce resource. Creating a farm that is not only productive now but also productive in the future is the only way of ensuring that the business in question does have a future. Additionally, if the landowners themselves lead the way and make the necessary changes that will benefit themselves and others, then there will be less likelihood that the changes will be imposed on them via new legislation. A self-regulating industry is just one of the many beneficial spin-offs of pursuing sustainability.

Production planning

Economy of scale

The size of a farming operation can dictate the choice of what is produced. A good example is wheat. The large scale at which wheat is currently produced limits the price per tonne paid, making wheat production for the open market unprofitable for smaller operations. The producer who wants to grow wheat but cannot do so on a large scale should investigate markets for products such as organically grown wheat. A greater effort is required to produce organically but the return is greater too, making it a viable option for smaller wheat growers. These niche markets can be very profitable as there is a market demand and smaller supply of the product.

In assessing the scale of economy for production, estimate the amount of production for the area to be used, including the highest and lowest production expected. Then estimate the amount of return expected versus the cost of production expected at these different possible production rates. There will always be a point where the cost of production of each unit (ie tonne of product, number of livestock, etc) starts to rise higher in relation to the return. For example, when calculating the return on sheep reared for meat production, the cost of each sheep raised will decrease per hectare up to the point where the hectare can no longer feed any other sheep without additional feed being supplied. Once additional feed needs to be supplied, the cost of production will rise, so the amount of return for each sheep will start to decrease. This sort of information will help you to make an informed decision as to what can be included on the farm on an economical scale.

Keep in mind that most of the figures you come up with will be subjective. Finding out exactly what a market cost will be at a given time, or what actual level of production you will achieve is virtually impossible. However, this exercise is a valuable decision making tool, allowing the farmer to see what scale of economy can be achieved, what can be included in the overall production plan, and at what levels. It is difficult to overestimate the importance of research and forward planning to a successful farming operation.

Materials

Another important point to consider is the longevity of the product you are considering producing, especially in terms of materials. Is the product obscure, requiring sourcing of seed, special fertilisers, etc. from only one or two suppliers, or even overseas? Or is it a commonly produced item, with materials and the necessary related supplies being readily available at a competitive price? While being one of a select few producers is ideal, especially if there is a high demand for the product, assessing the ongoing supply availability is important. If you are entering a new area, where suppliers are few, try to investigate the supplier's reputation and keep aware of how they are doing. Risk brings high return, but a sudden loss of supplier can leave the producer without a product, and in financial trouble.

Equipment

Investment in proper equipment is costly, but very necessary. It is not essential to have the very best and newest of everything but investment in good, safe, up to date equipment can make the difference to efficient production. Some points to remember are:

- Upgrade equipment when necessary – to remain competitive, you must use equipment that will enable you to produce competitively (eg if your competition is using machines that cut the cost of production significantly and you are not, you may be squeezed out of the market).
- Foresee the lifespan of machines/tools and buy according to that lifespan. If a machine is not likely to be outmoded, buy a quality model, maintain it well, and sustain its use for a long time to maximise benefit from that equipment. Look at the depreciation value for tax deductions as well.

Figure 7.1 Expensive machinery like this nut harvester may need to be purchased when diversifying or changing farm production. It is important to plan ahead and realise such costs before changing production.

Value adding

This book covers sustainable farming methods that can be employed to ensure the long term productivity of available resources. One aspect that needs to be considered is the maintenance of economic viability or economic sustainability. Value adding is one way to ensure the ongoing financial stability of a sustainable farming enterprise.

What is value adding?

Value adding refers to the development of extra farming income or profits through judicious assessment of current profits, available assets and awareness of markets and marketing trends. Many primary producers are being forced to re-evaluate their farms in order to stay competitive. By developing a product they already produce or by increasing the quality of their produce, they can open up new markets and yield greater profits. Examples are the production of specialty produce such as fine wool, goat's/sheep milk made into marketable dairy products, bed and breakfast accommodation, horseriding facilities or similar attractions for visitors to the farm.

Figure 7.2 Tourist activities such as these camel rides can be a source of income for farmers located near tourist centres.

There are many avenues for value adding on any farm. What is required is a common-sense approach that identifies possible areas for development backed up by market research of potential customers and practical marketing and advertising. Bed and breakfast, for instance, seems a good option for value adding, but is only really feasible in an area that already has tourism developed to attract guests.

Examples of value adding concepts:

- producing wine instead of selling wine grapes
- sun-dried tomatoes bottled and marketed to shops instead of fresh tomatoes
- producing dried flowers instead of marketing fresh cut flowers
- growing Asian specialty vegetables and herbs that are packaged and sold directly to retailers

Value adding is basically taking the primary product one step further. It requires more investment in the product, but the return is also higher. Rather than selling the produce to a middleman, who will then reap the rewards of processing the product, usually at relatively low cost, the end product is processed by the producer.

Value adding has been the starting point for many cottage industries and even the entry into overseas markets for many producers. Another advantage of value adding is that the producer is no longer at the mercy of fluctuating market prices for produce. If the price for a primary product falls, then the cost of producing a value-added product also falls. The producer who value adds is more likely to move through market fluctuations with greater ease than the primary producer.

If the sustainable farm is producing a variety of products, as it should be, it would be difficult to pursue value adding for the entire range of produce. However, a specialisation in one area in regard to value adding can be economically pursued.

Organic certification schemes

The organics movement is rapidly gaining popularity as consumers look for healthier, safer food and fibre products. Industry response, both in Australia and overseas, has been to establish a number of organic certification bodies to carry out quality assurance monitoring. In Australia, there are three main certification bodies – the Biological Farmers Association (BFA), the Bio-Dynamic Agricultural Association of Australia (identified by the Demeter trademark), and the National Association of Sustainable Agriculture Australia (NASAA). There are equivalent bodies overseas, working with the umbrella organisation, the IFOAM (International Federation of Organic Agriculture Movements).

Certification is obtained through on-farm and produce inspections to ensure the product is what it claims to be. The produce is graded according to its organic status (eg NASAA has three grades: A: Organic; B: Conversion to organic; C: not organic).

The certification bodies work with government agencies and equivalent bodies overseas to ensure that organic produce complies with national and international standards. One of the main aims of the standards is to restrict what is being sold as 'organic' and to provide more universal grading and packaging systems to reduce confusion amongst consumers.

It is in the organic grower's interest to obtain certification. Within Australia, all major outlets, such as supermarkets, and all organic wholesalers around the country will deal only in certified produce. While it is not illegal to sell uncertified produce as 'organic', it is illegal to sell it as 'certified organic'. Exported organic produce is subject to more stringent rules. Any product carrying an organic or biodynamic label must be accredited by an AQIS-accredited (Australian Quarantine Inspection Service) organic certifier such as the BFA or NASAA, and they must comply with the standards set down by the IFOAM. In many other countries, similar certification programs exist.

The decision for a farm to become organic is not an overnight process. Although it can be started at any time, organic farming requires much planning and forethought. Changeover tends to be a transitional process that takes into consideration the state of the farm in its present condition and its suitability to organic farming methods. The type of organic classification the farm is pursuing (ie organic, biodynamic) will dictate what procedures, tests and methods must be used, and what time is involved. Contact your local or national certification board to obtain guidelines on what is involved in converting all or part of a property to organic growing.

Contingencies and seasonal variations

Planning is a continual part of any business. While pre-planning can help to prevent a lot of unexpected difficulties, there are always things that cannot be foreseen. Nothing ever remains the same in agriculture, especially in regard to seasonal changes. Look at what can be expected and what may be considered 'the unexpected'. Action plans in these areas will not prevent the event from happening, but they can stop events from becoming catastrophic.

The expected

Seasonal variations are a fact of life. Many years of good growing weather may be followed by a few years of drought or heavy rains. This is one of the benefits of choosing to diversify. One product may have average results while others thrive in the changed conditions. The move away from a monoculture will ensure that the need to dispose of or undersell excess stock does not have a devastating impact on the producer. With diversified product, a loss in one area will often be offset by a gain in another.

Another consideration is the timing of the produce and its effect on the farm and the farmer. How and when something is produced can affect the soil. For example, heavy hoofed animals left in paddocks during wet seasons can severely damage the soil. The answer may be to house animals in a barn during the wet season. This is doubly beneficial if shelter is needed for drying, such as for herbs, during the hot season. The building is in constant use and the soil is spared from the worst aspect of animal impact, making it markedly more sustainable to the system.

When crops are to be planted and harvested should also be considered. It is of no use to have several different crops, only to find that they are at their peak harvest at the same time of year. Extra help may be needed to harvest the crops, bringing additional costs of production. While everyone needs a 'slow' period in which to regain their sanity, the staggering of planting times and production sees resources being used to their best advantage at all times.

The unexpected

In any agricultural situation, the unexpected will happen sooner or later. Droughts, floods, cyclones and pest plagues may be uncommon, but they DO happen. Without good forward planning, these events can easily send a farm bankrupt. On the other hand, with good planning based on informed foresight, you can minimise the impact of these otherwise potentially disastrous situations. A sustainable approach to farming is, in itself, a way of minimising the likelihood of problems.

Types of problems

- Drought – plants and animals dying; some native plants may go into a dormant phase if it is a natural occurrence in the region. Non-indigenous plants will be more likely to die. Consider producing 'bush tucker' (indigenous food plants) in a drought-prone area. There are more and more drought resistant animals appearing in drought prone areas. This is a good thing provided farmers don't overstock their paddocks simply because these animals can survive. This kind of overstocking may not immediately kill stock, but it will certainly degrade the land and reduce its carrying capacity.

- Flood – drowning of domestic and wild animals; erosion; crop destruction; structural damage to fences, buildings, roads etc. are all common results of flooding.
- Storm – strong winds may damage buildings and stock fences and blow over crops and windbreaks; torrential rain may cause erosion and landslides; hail can strip crops of their foliage.
- Plague – pest and disease can wipe out crops and even destroy biodiversity of natural bushland; animal plagues may result in either quarantine restrictions or stock death.
- Soil degradation – there are many and varied causes (erosion, storms, floods, droughts, vehicular traffic, hoofed animals, removal of vegetation, etc.).

Solutions

- Reduce production so as to not overload the property, eg fewer animals per acre
- Diversification of produce
- Consider alternative types of produce, eg use alpacas rather than sheep because they do less damage to soil
- Adjustment of production for cyclic crops
- Harvest earlier/later, eg in a cool season, harvest wine grapes later because it takes longer for sugar levels to develop; if a storm is coming, harvest earlier to avoid damage; in warm areas consider early cropping fruit trees to avoid fruit fly infestation.
- Postpone planting – in harsh weather conditions postpone till the outlook is better. Delaying planting will also result in later maturing crops which may result in gaining a stronger marketplace when other stocks of crops are depleted.
- Better landcare – improve carrying capacity of a property (sometimes simply by adding a deficient micronutrient).
- Minimise use of heavy equipment to lessen land damage.
- Consider things such as permanently flooding a frequently flooded area for production of rice, etc.
- Ensure there is a high and dry area for livestock commonly kept in flood zone areas.

Planning for drought

Some localities are more susceptible to drought than others. The risk of drought in any area needs to be recognised and farming practices determined accordingly.

The effect of a drought is not only a reduction in farm capability but, as a result of this, an increased strain on financial resources, increased susceptibility to land degradation, etc.

Continuous monitoring and review can greatly help foresee the onset of a drought earlier than it might otherwise be detected. However, the signs are generally subtle and can easily be undetected until the farm is actually in drought.

Things to do before a drought *(all the time):*

- Regularly monitor water supply (quantity and quality)
- Monitor weather forecasts and trends (both short and long-term)
- Monitor feed supplies and costs in the market place
- Maintain a backup store of feed
- Consider likely future growth and value of a pasture when determining the degree to which that pasture will be grazed (eg when feed values are low, a pasture can be grazed more heavily, but when feed values are rising, the pasture should be grazed less)
- Manage the total farm to maintain quality with tree plantings, water sources, etc. (this reduces erosion impact of drought and helps recovery)
- Retain enterprise diversity and flexibility so that the mix can be changed with relative ease to something more appropriate during drought
- Diversify sources of income (both on and off farm) so you have an income during drought (eg off farm investments, value adding, ecotourism)

Things to do during a drought*:*

- Progressively reduce grazing/cropping pressure (eg provide supplementary feed for stock, and restrict access of stock to susceptible pastures; reduce number of crops grown – instead of two crops a year on a paddock, only grow one)
- If drought extends over a long period, some farming enterprises may need to be curtailed or changed
- Reduce stock (eg sell stock at market, send to agistment elsewhere)
- Maintain breeding stock (may need to send away for agistment; may need to provide supplementary feeding etc)
- Retain stubble/crop residue on all areas for erosion control
- Watch finances more closely and don't live on hope – act while you still have something in reserve, even if it means stopping farm production and seeking off-farm employment temporarily
- Talk to banks/financial advisers before implementing any major changes

After the drought

The transition from drought back to normal conditions must be managed carefully. It can be full of risks. Drought-breaking weather can be dramatic and, as such, can have serious negative impacts on animals, plants and soils.

- If the soil surface is dry and bare it can be easily eroded by rain
- If animals are weaker than normal, a cold snap or wet weather may result in disease or infection
- Dramatic changes in feed can cause digestive upsets in stock (they can overeat; they may take in unmanageable quantities of plant toxins)
- There must be a transition from poor quality drought-affected pasture to lush new pasture following rains (perhaps limit stock to new pasture for short periods of an hour or two daily – increasing gradually)
- Pests, diseases or weeds may develop rapidly when the drought breaks
- The drought may have left livestock unhealthy and deficient in nutrients, requiring special attention to recondition them

- Soils may be degraded (eg nutrient or organic content deficient) and require treatment before planting a new crop
- Certain pasture species may have disappeared, so it may be necessary to reconstitute a desirable pasture mix (often legumes do not survive as well as grasses so they may need to be replanted)
- There is a temptation to spend money (even through borrowing) to rebuild the farm quickly but this is not necessarily the best way; compare the cost and benefits involved in a 'quick fix' (eg buying livestock) to a 'slow fix' (eg breeding up livestock numbers)

Excessive water

Excessive water does not apply only to heavy flooding. Over-wet soils will slow the growth of most pastures or crops. Excessive water in the soil will displace air in the soil, starving the plant roots of oxygen. Plants growing in over-wet soils can begin to exhibit nutrient deficiencies. Usually the first signs are a yellowing of foliage (particularly older leaves), caused by an inability to absorb nitrogen under wet conditions.

Abnormally excessive moisture may be caused by unseasonal heavy rains, flooding and deteriorating (or changed) drainage conditions.

Once a soil passes being damp, and becomes excessively wet, the soil structure also becomes more sensitive to damage. Grazing on an over-wet soil can damage soil. Animals walking over wet pasture can destroy up to 50% of otherwise useable biomass (pasture mass). Pasture also tends to be spoilt by animal urine and manure more easily in wet and windy conditions. Pugging of soil can occur – bare patches may develop.

Take extra care to protect over-wet soil from livestock. For example, if cattle are provided with excessive feed, they tend to eat heavily early in the day, then spend the rest of the day walking about, damaging the soil as they burn up the energy they consumed. Offer them only around 70% of their daily needs (normally consumed the first three to four hours of the day) during the day and move them to another paddock to offer them the other one-third at night. This will tend to reduce damange such as pugging, as well as hoof damage to the soil. If a property gets too wet, it may be cheaper to put cattle out to agistment than to allow damage to the soil and pay the cost of repairing it.

What other planning do I need?

Planning is an ongoing activity. Most businesses should prepare, at the minimum, one-year, five-year and 20-year plans. Included in the business plan are the planned production rates, estimated costs and expected returns. There should also be plans for investments, equipment needed and possible new areas to pursue, and, as outlined above, plans for unseasonal weather and catastrophic events.

All plans, though, should be subjective, ever changing and adapting to what is happening in the real world. For instance, many producers who have turned to value-added products are reaching their 20-year goal in five to ten years. Without a basic plan to guide and help the business keep its 'eye on the ball', changes can mean the end of the business instead of the breakthrough to the future. Planning is the best way of being as prepared as anyone can be for whatever surprises the future may hold.

8

Managing plants – Crops and pastures

Whether you produce animals or harvest plants, the basis of any farm is still its plants. For a farm to remain sustainable, certain minimum productivity levels must be maintained, using preferred plant species on an ongoing basis. These plants may be pasture species, fodder crops, grain, vegetables, fruit or other harvested plants. This chapter shows you some of the techniques that are important to ensure sustainability of plant growth.

Selection criteria for plants

- What crops are currently in demand? You need to attempt to gauge future demand, particularly if you are looking at growing crops that are long-term investments and may take several or more years to reach a marketable stage (eg tree fruits, nuts, timber). Also look at the 'stage' of demand for a crop. Is it a new, growing market, or is it one that everyone is 'getting into' (resulting in a possible glut on the future market)? Select crops that are in high demand, where possible, to remain economically sustainable.
- Which crops are suited to growing in your locality? Some alteration to the soil and climate of the area may be beneficial in the long term. Examples are the introduction of windbreaks to prevent erosion, installing irrigation systems, or the creation of a microclimate to encourage growth of a particularly suitable plant.
- What resources do you have to produce different crops? This could include suitable land, equipment, staff, materials, or the financial backing to obtain these. Investment in equipment and materials must also be balanced with the amount of return you can expect.
- What expertise or knowledge do you have with regard to growing different crops? Can you obtain that knowledge? For new or experimental crops, determine what

information is available on their culture and find out what grower support exists (eg department of agriculture). Trying crops new to your area or still in an experimental usage stage can be costly but it has the potential to be very rewarding. Overseas research can often shed light on the suitability of the crop for your area. Start small and work up to larger production numbers if the results are good.

- How will the crop under consideration work with other crops? For instance, is there a market for a suitable companion plant? What crops should it be rotated with? What effects will this have on the soil and on the economics of growing this plant? Can the crop be marketed easily in conjunction with other crops you produce?
- What will you be using the crop for? If you are considering crops for your own subsistence, is this the cheapest and easiest way to obtain the crop? If you are using it for stock feed, is this the cheapest or easiest way to obtain suitable stock feed?
- Is the crop sustainable? Many crops can only be grown with large inputs of fertilisers and pesticides. Choose crops that are suitable for your soils and the surrounding ecology.

Grain and other broadacre crops

Monoculture

Monoculture is the most prevalent form of production in Western agriculture today. It refers to a system of growing large areas of a single crop in which almost no diversity is present at all. Crops grown in this way are often especially open to attack from weed and pest species. Many predators return annually to these farms, assured of a continual food source. The stripping of crop-targeted nutrients from the soil is also a major problem in a monoculture. To combat these effects farmers are required to use greater quantities of chemicals in the form of weedicides, pesticides and fertilisers.

Classic examples of monoculture can be witnessed throughout continents such as Australia and North America where vast tracts, millions upon millions of hectares of land, are used for wheat and other grain crops. The species being produced are generally fast growing, high yielding, hybrid varieties requiring considerable chemical inputs. They are often sterile varieties and seed must be purchased for each planting. The seed suppliers are often the same or sister operations to those that provide the required chemicals needed to protect the crops from weeds, insects and disease.

Aside from the problems of poor land management and heavy use of chemicals that the monoculture farm can create, the primary producer must remain viable. Quantity of production and most productive use of land can be heavily influenced by perceptions of economic viability.

There are examples of predominantly monoculture systems that are relatively successful in terms of sustainability. The reason for this is because the people who use these systems are aware of the dangers of monoculture, especially in terms of chemical use, and have therefore developed sustainable natural defensive measures.

One method that is employed is to plant species-rich 'islands' at intervals throughout the crop. These resource islands, which can be made up of literally hundreds of different indigenous plant species, seem to work quite effectively at controlling pest and disease populations as well as increasing soil fertility.

Research is still being conducted to assess to what degree these islands are successful but it would appear that the concept works. Further work to determine which alternative species are the most beneficial will help to ensure the resource islands are most effective. This concept is very similar to the permaculture ethic of companion planting although it exists on a far grander, and perhaps greater, scale of diversity.

Crop rotation

Many of the problems associated with monocultures can be minimised simply by rotating crops. As a general rule, in situations where there are more problems, leave greater time periods between plantings of the same crop. Sustainability may be improved by the following:

- Grow a crop or crops for half of the year, and graze the same area the other half
- Grow several different crops on the farm, and rotate them so the same crop is not grown in the same paddock more than once every two to three years (or preferably longer)
- Fallow areas between crops (ie do not graze or grow a crop during the rest period)
- Grow cover crops for green manure at least annually to revitalise the soil
- Ley farming systems – this involves alternating cereal grain production with pasture Annual medics or sub clover, mixed with grasses, are useful to produce high quality forage.

Row crops

Row crops may include such products as maize, vegetables, cut flowers, herbs and berries. They are often, but not always, replanted periodically. As such, the ground needs to be cultivated and a “seed bed” prepared. Poor seed germination is the result if the soil is not prepared yet cultivation, especially of large areas, can cause major problems with erosion.

The following techniques will not only help to control erosion, but will also make row crops desirable in a sustainable agriculture system. They include:

- Inter planting temporary crops (eg vegetables and other annual plants) with permanent crops (eg fruit trees or vines). Another option is to grow grass or other ground stabilising ground covers between rows of permanent fruit trees, vines or flower crops (eg woody perennial flowers). Cover crops that can be tilled into the soil to enhance its properties are ideal in these situations, providing they are not competing too much with the main crop for soil and nutrients. Growing low-growing legumes such as clover between rows can add valuable nitrogen to the soil.
- Restrict row crop length on steep slopes (perhaps to 70–80 m in higher rainfall areas) to minimise runoff effects. Rows should be positioned across the slope. Rows running down the slope will encourage erosion.
- Maintain a grass strip at the end of each row to catch runoff of water and soil particles.
- Use a large tine to deep rip areas where a tractor has worked (and caused compaction) to increase the depth of water penetration.
- If using a plastic mulch is necessary (eg in strawberries), water runoff is increased, so areas between rows need to be planted (eg with ryegrass, barley or clover). The second crop can be harvested or used as a soil enhancer.

Cover crops

A cover crop is simply a plant that is grown for the purpose of improving the condition of the soil in which it is grown. It is most commonly ploughed in, but can also be cut and left to lie on the soil. The latter method is very slow, but can be effective.

In theory, a cover crop should increase organic content and fertility of the soil, but research has shown that this is not always the case. The real contribution of a cover crop is affected by:

- The amount of growth achieved
- The plant varieties grown (eg legumes add more nitrogen to the soil than they take out)
- Whether any part of the cover crop is harvested and removed from the paddock (perhaps as hay)
- Whether there is a strong leaching effect (eg in sandy soils or on steep slopes)
- Temperature and moisture conditions – excessive heat and moisture can result in rapid decay of organic material and little, if any, increase in soil organic content; excessive dryness can result in very little decomposition
- Carbon: Nitrogen ratios of residues – (high ratios such as 100:1 are slow to decompose but lower ratios may be much better)
- Soil life – the presence of certain micro organisms, worms, etc. can have a significant bearing upon decomposition, release of nutrients and even mixing of residues into the soil mass.

A recent survey of farmers in north-eastern USA found that farmers were using cover crops for varying combinations to:

- Improve soil fertility, soil structure or tilth
- Control erosion
- Reduce the need for fertiliser and other soil amendments
- Increase nitrogen levels (ie legumes as a green manure)
- Improve nutrient availability
- Minimise leaching
- Weed, pest or disease control
- Prepare land for production of other crops (eg vegetables or grain)
- Use as a livestock feed supplement

The cover crops used must be matched with the desired outcome.

Cover crop guidelines/principles

The following tips will help in determining selection of a cover crop:

- Type of crop – perennial crops are generally preferred over annuals; with annuals, large populations of nematodes often move into the soil after maturing, causing problems for the root system of any subsequent plantings
- Effect on soil pH – alkaline-tolerant plants such as sorghum and barley can be grown to reclaim alkaline (lime) soils. Growing a single crop of these plants may cause sufficient acidification to allow less lime-tolerant legumes to be grown, further acidifying the soil and allowing it to be used for livestock or a cash crop.

- Timing – the crop should be incorporated (tilled) before maturity (ie before flowers and seeds form)
- Water use – while cover crops, like any other crops, do use water, their root growth can lead to better penetration of water into the soil; additionally, residual organic material left by the plants will lead to increased water conservation

Legume cover crops

Legumes commonly have 15–30% more protein than grasses, giving them better food value for livestock. Another advantage of legumes as a cover crop is the production of rhizobium. Rhizobium is a bacteria with which legumes can be inoculated, resulting in production of hydronium ions in the soil. These ions in turn lower the soil pH, increasing its acidity.

The decomposition of organic residue also has an acidifying effect on soil. Increased organic matter does however buffer (ie tend to slow down) this acidification. Nevertheless, excessive and continual use of cover crops, especially legumes, without liming or use of a similar treatment, can result in soil becoming too acid and losing productive capacity.

Inoculation of legumes

You can use pre-inoculated or pelleted seed, or you can inoculate seed yourself.

Inoculating seed

- Add the inoculant to another medium (eg peat mixed with water and gum arabic) – use 1 part sticking substance (eg gum arabic) to 10 parts water; other sticking materials that can be used include corn syrup, sugar, powdered milk or various commercial stickers
- It is critical to use only fresh inoculant in the appropriate concentration
- Use the appropriate rhizobium for the legume being grown; keep in mind that rhizobia perform better on some legumes (eg alfalfa) when seed is coated with calcium carbonate, while others perform better when left uncoated (eg red clover)
- Check the expiry date – commercially produced, pelleted seed should be sown as soon as possible; at least within four weeks of production, as it does not store well
- Always store inoculant in a cool, dark place
- In dry conditions, inoculant rate may need to be doubled
- If legumes exhibit yellowing of foliage, this may indicate nitrogen deficiency resulting from failure of the inoculant
- Applying some nitrogenous fertiliser when planting a cover crop may actually enhance the nitrogen fixation of the legumes (eg around 30 kg per hectare of starter nitrogen)
- Generally soil pH needs to be over 5.5 for rhizobia to survive

Shade-tolerant cover crops

These include cowpea, burr medic and hyacinth bean

Salt-tolerant cover crops

Strawberry clover, white clover, burr medic, field pea, barley 'Salina', are all ideal for use in areas of high salination or heavy salt spray.

Types of cover crops

Alfalfa – see Lucerne

Barley

Growing conditions

- suited to cool, dry climates, including higher altitudes
- moderate frost resistance
- moderate drought tolerance

Soils and nutrition

- will tolerate alkalinity but not highly acidic conditions
- high tolerance of salinity, best of the cereal crops
- moderate biomass production as cover crop
- grown to increase organic content of soil
- strong, well established root system aids erosion control

Uses

- hay, grain and silage
- good cover crop and green manure properties prior to cash crop sowing
- light grazing potential
- used in conjunction with other cover crops to reduce weed infestations
- improves water infiltration rates

Problems

- host for Thompson seedless grape nematodes
- not as suited to companion planting as some cereal species because of competitiveness

Buckwheat

Growing conditions

- warm season crop, plant late spring in temperate areas
- plant seed at 30–45 kg per hectare

Soils and nutrition

- tolerates poor soils

Uses

- to smother weeds when densely planted (fast growing)
- as a green manure cultivated in seven to ten days after flowering (around five to six weeks from planting)
- deep rooting and can increase nutrient availability and improve soil structure

Problems

- frost sensitive
- can harbour root nematodes

Canola

Brassica napus (also known as rape or rapeseed)

Growing conditions

- plant seed 2.5 mm deep
- sown in spring or autumn in temperate areas

Soils and nutrition

- has the ability to accumulate nitrogen that otherwise may be leached from the soil (better than most other non legume cover crops)

Uses

- grown to suppress weeds, increase organic content and encourage soil life
- decomposes fast when tilled into the soil
- attracts various types of hoverflies which are predators of aphids
- grown as a cover crop, forage plant, for bird seed, or to produce foods (eg canola oil, used in cooking, for margarine)

Problems

- avoid growing in areas where brassica crops have been grown, or brassica weeds (eg wild radish) are growing; as this can lead to build up of pests such as aphis.

Field Pea

Growing conditions

- requires a reasonable tilth and even seed bed
- sow in autumn, grow over winter; at 2.5 to 7 mm deep
- germinates quickly
- germinates at temperatures as low as 5°C, although germination is better around 24°C
- does not tolerate excessive heat, dry or wet conditions
- intolerant of salinity
- does not self seed very well; requires replanting
- has been grown successfully in semi-shade between nut trees
- does not regenerate well after mowing

Soils and nutrition

- prefers reasonably fertile, drained soils
- grows in pH from 4.2 to 8.3; it has greater ability to acidify soil than some other legumes (eg lupins)

Uses

- a cover crop rotated with vegetables or field crops; excellent for raising nitrogen
- used for forage, hay, silage, grain or green manure
- useful for weed competition in areas with strong winter weed growth (roots exude a chemical that inhibits seedling growth of some grasses and lettuce)
- suppresses weeds better when inter planted in high density with barley

Problems

- susceptible to various pests and diseases (Fusarium, Sclerotinia, powdery mildew, aphids – so avoid preceding or following with other plants susceptible to such problems)
- excessive use can cause acidification
- not tolerant of extreme conditions
- susceptible to various nematodes

Lucerne

(*Medicago sativa*) (also known as alfalfa)

Growing conditions

- for better establishment lucerne should be grown with a companion grass such as perennial ryegrass
- for maximum production lucerne should be cut for hay as any grazing will reduce yields significantly; loss of leaf should be kept to a minimum by careful handling
- in dryland areas lucerne is used only for grazing; yields are kept high by use of rotational grazing practices with at least two spells of rest, each of four to six weeks duration, through the growing season
- will grow at medium to high temperatures if there is sufficient water and humidity is not high

Soils and nutrition

- adaptable, preferring moist, but well drained soils
- deep-rooted, so will tolerate dry periods
- prefers neutral to alkaline soils
- responds well to superphosphate applications

Uses

- highly palatable, productive fodder crop
- one of the best legumes for raising nitrogen levels in soils
- useful for outcompeting some problem weeds

Problems

- heavy user of calcium and repeated cropping of an area can result in increasing acidity; regular liming required for repeat cropping
- danger of bloat when grazed by ruminants; lucerne for grazing is best mixed with other pasture species
- some varieties prone to pest problems, such as aphids, red legged earth mite and lucerne flea, and to diseases such as verticillium wilt
- lucerne is difficult to make into silage, unless it is grown with a companion grass crop, as the high levels of alkaline materials contained in the plant tend to the neutralise the acid levels required for good fermentation, and make the silage unpalatable.

Lupins

(Including *Lupinus albus*, *L. luteus* and *L. angustifolius*)

Growing conditions

- generally suited to cooler climates and often grown as a winter annual

Soils and nutrition

- suitable for a wide range of soil types, with some species tolerating saline conditions
- moderate nitrogen fixing qualities,
- positive soil improvement capabilities such as aeration and opening of compacted soils due to deep taproots
- increase availability of phosphorus, manganese and nitrogen to surrounding plants, making it ideal for intercropping with cereals such as wheat or oats
- beneficial insect-attracting qualities
- gradual lowering of pH in alkaline soils

Uses

- alkaloid free lupins used for silage
- alkaloid present lupins used as cover crops in paddocks out of stock rotation

Problems

- possibility of poisoning in livestock due to quinolizidine alkaloids
- evidence of harbouring of pest insects in low alkaloid strains of lupins
- some intolerance to even low level herbicides in some species of lupins

Oats

(*Avena sativa*)

Growing conditions

- cool season grass growing to 1.2 m tall in temperate climates
- seedlings may tolerate low temperatures (to –8°C)
- susceptible to hot dry conditions
- tolerates wetter conditions than barley; needs more moisture than many other small grains
- typically seeded at 90 kg per hectare
- sow 2.5–5 cm deep

Soils and nutrition

- grow well on wide pH range (tolerate to pH 4.5)
- grow on wider range of soils than most other cover crops, fertile or infertile, sandy or not sandy
- for maximum yields, NPK fertiliser application may be needed at planting (require nitrogen fertiliser, in particular when temperatures are low)
- less salt tolerant than barley

Uses

- hay, pasture, green manure, cover crop
- very palatable to livestock (more than cereal rye)
- in North America, grown in rotation with corn
- grown for silage and hay prior to seed maturing
- oat straw is excellent for animal bedding
- in some places sown in late summer/early autumn as a cover crop, following a cash crop
- high C/N ratio so decomposes slowly

Problems

- easily overgrazed because of high palatability to livestock
- in most countries, oats are highly susceptible to a wide range of diseases and pests but various resistant varieties are now available

Ryegrass

(*Lolium multiflorum*, annual or Italian ryegrass)
(*Lolium perenne*, perennial ryegrass)
There are also hybrids and other species.

Ryegrasses are the most important pasture grasses throughout the world. They are tuft forming; and many varieties with varying characteristics are available.

Growing conditions

- generally prefer mild and moist conditions
- frost resistant
- Italian rye is a cool season crop

Soils and nutrition

- grow best on fertile soils
- selected varieties adapt to varying soil conditions

Uses

- Italian ryegrass is used for temporary pasture
- Perennial ryegrass is best suited to permanent pasture in areas with dependable rainfall

Problems

- Drought

Sorghum

There are hundreds of cultivars but they vary in characteristics and uses. They may belong to any of a number of species, the most common being *S. bicolor*. Some authorities divide them into four main groups:

1. Grain sorghum – non saccharine plants, grown mainly as grain for livestock; similar nutrition to corn but higher in protein and lower in fat; most have a relatively dry stalk

2 Sweet sorghum – stalks contain more sugar, used for forage, silage or making molasses
3 Broom corn – stalks are very dry and woody; grown for making straw brooms
4 Grass sorghum – grown for pasture, silage, hay, cover crop; until plants are at least 50 cm tall, prussic acid in foliage can cause food poisoning to grazing livestock

Growing conditions

- warm season annual grass, up to 2 m tall
- sow after threat of frost in spring
- frost tender
- sow 2.5 mm deep in moist soil or 5 cm deep in dry soil
- rows spaced at 25–50 cm, seeds 5–10 cm apart
- germinates ideally at soil temperature of 18°C
- tolerates high pH, may be used with barley to reclaim alkaline soils
- needs minimum annual rainfall of 400 mm; preferably higher

Soils and nutrition

- best in reasonably fertile and friable soil, but will adapt to many other soils
- uses high levels of nutrients, so for optimum results, farmers may apply up to 160 kg of nitrogen fertiliser per hectare on poor soil (half or less on fertile soils); phosphorus and potassium are often not needed

Uses

- uses vary according to type (see above), though any forage types are good as a cover crop, to increase organic content, promote microbes and control weeds
- performs well as a cover crop mixed with cow pea, buckwheat or sun hemp
- a crop can be cut and baled, three or four times (at 60-day intervals) over a season
- very high C/N ratio, so slow to decompose

Problems

- crops usually fail because of cold soils, hard soils (soil crusting or poorly prepared seed bed), poor seed quality or incorrect planting depth or spacing
- young plants may be toxic to livestock
- can harbour nematodes that may lead to reduction in productivity of vegetable crops following sorghum
- as a cover crop, it can lead to a decrease in nitrogen availability
- can harbour pests of some other plants (including pecans, and more particularly cereal and grain crops); NB: many new varieties have disease and insect resistance bred into them

Subterranean clover

Growing conditions

- self seeding, easily established annual
- prefers cool or mild winter (dies in heat)
- early maturing varieties planted autumn to mature in winter
- water requirement varies according to variety but generally 400 mm or greater

- if grown with white clover, white clover dominates in wet soil and sub clover dominates in dry soil
- mowing and grazing help control weeds in sub clover

Soils and nutrition

- does not do well on alkaline soils
- tolerates pH 5–8, prefers 6–7

Uses

- the most useful annual clover
- as green manure; component in high quality pasture, weed suppression and nitrogen fixation, useful in orchards; (NB: there is some evidence that sub clover may cause a reduction in grape productivity, though the mechanism is unexplained. This does not appear to be a problem with white clover though)
- many authorities claim that benefits are maximised when grown mixed with warm season grasses; however, shading by grasses can weaken sub clover
- some claim overall clover may be more productive when sub clover is mixed with another clover species such as crimson clover

Problems

- nematodes can develop and may affect succeeding vegetable crops, but the true significance is unknown

Trifolium (clovers)

Growing conditions

- usually a temperate climate species
- clover requires constant close grazing in order to control weed and other competitive pasture species

Soils and nutrition

- an excellent nitrogen fixing species
- pH range 5–10.5
- requires suitable amounts of phosphorus
- some species of *Trifolium* are saline tolerant

Uses

- high stock production properties, especially in dairy stock
- encourage microbial soil activity
- increase water infiltration and holding capacity
- controls soil erosion through heavy root system
- can be used as hay and silage

Problems

- cattle that are exposed to lush new clover growth should be drenched to avoid bloat
- several nematodes species can cause damage to *Trifolium* pastures
- periodic local damage can result from numerous insect pests

White clover (*Trifolium repens*)

This is a persistent, perennial legume and many different cultivars are available. Some types can grow to 25 mm tall.

Growing conditions

- small leaved forms generally tolerate a wider range of conditions than large leaved forms
- grows best under cool, moist conditions
- varieties bred to tolerate different conditions (eg poor drainage, drought, heavy soil, acidity, salinity, alkalinity, etc.)
- less heat tolerant than strawberry clover
- more shade tolerant than strawberry clover
- seed sown at 0.4–5 kg per hectare
- responds well under grazing or mowing
- often sown with barley or oats in autumn (these plants establish faster than clover and help nurse the clover until it becomes established)

Soils and nutrition

- generally best on well drained, fertile loam or clay soil
- most cultivars do not tolerate high salinity
- grows under pH 4.5–8.2; ideally pH 6–6.5

Uses

- one of the most nutritious forage legumes, often used as an irrigated pasture plant
- grown under some fruit orchards and vineyards in the USA (sometimes mixed with strawberry clover, birdsfoot trefoil and red creeping fescue)
- probably better suited to vineyards in particular, than subterranean clover which may inhibit grape production
- creeping habit is excellent for soil stabilisation
- because it dries slowly, if harvested, white clover is better used for silage rather than hay

Problems

- nematodes can damage white clover in some parts of the world

Other cover crop plants

Other plants which are often used as cover crops in different parts of the world include: annual fescue, barrel medic, burr medic, cereal rye, common vetch, cowpea, crimson clover, Kentucky bluegrass, millet, mustards, strawberry clover.

Ways of using a cover crop

1. The main crop in the primary growing season: grown in a paddock during a fallow year.
2. A companion crop: grown for its ability to repel insects, enhance flavour, or give other desirable benefits to the main crop.

3 A 'catch crop': grown between rows of a main crop, as a ground covering, controlling erosion and keeping the ground cooler. It can also be planted after harvest to catch nutrients and reduce leaching.
4 A feeder crop: grown amongst other crop plants to increase or maintain nutrient levels, eg clover grown amongst other plants helps maintain nitrogen levels in the soil for the other plants. Garlic and other related plants may raise sulphur availability to adjacent plants, increasing resistance to diseases.
5 An off-season crop: grown during a part of the year when the main crop cannot be grown.

Source: *Sustainable practices for vegetable production in the south*, by Dr Mary Peet, North Carolina State University.

Hay and silage

These are two different methods of storing/preserving harvested fodder. Hay is harvested dry, so quality haymaking relies on good weather. Silage can be made under poorer weather conditions, however fine weather is preferred.

Silage production

Silage involves harvesting fodder as a green crop and storing it in an airtight situation to minimise loss of nutrients. Silage is a fermented food source and it is this fermentation process that must be managed correctly in order to achieve success with silage. The fermentation process requires correct moisture levels, sugar content and pH levels. The process is in many ways similar to home pickling of onions or cabbage.

Harvesting

The type of silage or hay required will determine the growth stage at which the plants are to be harvested. It may be made from grass, legumes or other pasture or green manure species. Any of the following methods may be used for harvesting silage or hay.

Mower conditioner

This method can only really be applied to round bale silage in which good quality silage can be achieved. Power consumption costs are low, the method is relatively simple and output per hectare is both high and efficient.

Flail

Although reasonable quality silage can be obtained with a single chop or flail, the quality suffers due to bruising of the crop and variable lengths in the cut. Costs are low, but this method is really only suitable for small-scale silage production.

Double chop

This is a far more suitable method for silage production as the crop is far less variable in terms of chopped size of the product and therefore quality is increased. Power costs are a little higher but yield in terms of time and output per hectare is much improved over single chop.

Precision chop

High production and high quality offset higher running production costs. Not suitable to round bale, but the most intensive and results proven method for all other forms of silage.

The silage is, as suggested, very precise in chop size and this is an important factor in a consistent fermentation process.

Silage inoculants

There are a number of products that can be used as additives to aid the fermentation process. What these actually do is introduce a certain type of bacteria that dominates the fermentation and ensures speed of metabolism, and this contributes to a quality product.

Animals that are fed with silage that has been treated with inoculants benefit in production terms as an ongoing causal affect of the better quality feed.

Silage timing

Any fodder crop is suitable for use as silage. There are important timing considerations that will apply to the quality of the silage. Different crops yield higher nutritional qualities at different stages in their growth. Some examples of optimum silage production timing are:

Lucerne

The optimum time for using lucerne in silage is during the early flowering stage, although wilting is sometimes necessary in order to reduce moisture content. Too much moisture will slow down the fermentation process and result in poorer quality silage due to moisture runout taking valuable nutrients with it.

Pasture

Early flowering stage is usually correct with most pastures, as the plant is high in sugar content and almost at peak quantity size wise. Some pastures that are high moisture content varieties, clovers for example, will require that some wilting take place before chopping occurs.

Sorghum *(grain and sweet varieties)*

Sorghum has proven itself to be a premium dryland summer crop, being able to cope with little water and yet still give reasonable yields. Grain sorghum can be ratooned (regrowths from roots after cropping) but careful management is required in order to be successful. Sorghum is best harvested for silage at the milky dough stage and, like all grain crops, should be cracked in order to give quality feed.

Sorghum (forage varieties, Sugargraze, Jumbo, Pac 8260, etc)

These forage varieties of sorghum are high yielding, high quality, and also far more tolerant of ratooning for multiple crops without having to resow. They are best cut for use at a height of 1.0–1.5 m when they are at their peak in terms of quality and quantity.

Soybeans

During the early pod filling up stage is the best time for soybean.

Quality control

Good quality silage results from good quality forage crops that are cut at the correct time and then sealed quickly to start the fermentation process. Treatment with an inoculator is the next level in quality control and management. Soil contamination is a main source of poor quality silage and spoiling in the storage process. Care during the cutting and lifting stages will keep this problem in check.

The manner in which feeding out of silage occurs is another quality control factor, especially with the larger pit and bunker storage methods. Minimal disturbance of the silage face will mean that air will not infiltrate and cause excess spoiling further into the pit.

Storage and handling

There are varying storage and handling options associated with silage production and these tend to be related to the quality and quantity of the required end product.

Pits and bunkers

This is the traditional storage method of keeping silage and is still widely used even though it does have some limitations. The main advantage with this form of storage is cost. Spoilage, difficulty of handling and expected storage life are the disadvantages. The silage is either placed in a wedge-shaped pit or on a sloping, even surface and then covered with a heavy duty plastic or tarpaulins to seal air movement through the silage. Second hand tyres are often used to weigh down the covering material. Sometimes silage will be stored indoors and not immediately covered, in which case air flow through the building should be kept as low as possible.

Tower silos

These are structures that are specifically made for storing and feeding out. They are quite costly initially but are capable of high quality product with very little wastage. Dry matter content must be watched closely at the cropping stage and wilting is sometimes necessary to bring the moisture content down.

Roundbale

This is a method that involves producing compact, easy handled silage in bales. The silage is wrapped in a plastic covering which ensures its storage life. An important factor is the wrapping process, which, if not done correctly, will mean a downturn in product quality.

Silopress

This is similar to roundbale, but much larger in concept. The silage is pressed into long, thin plastic tubes that can have limited transport potential but are quite good in terms of self-feeding, costs and quality.

Hay

After cutting, the wind and sun are left to dry hay naturally. Dry weather is really needed for this period. During drying, the moisture content of the cut material can reduce from an original 80% to 20% or less. At this stage it can be stacked under cover. There are many different techniques for making hay, used in different parts of the world. During the drying period, the hay may be turned over with a machine, to allow better air circulation and faster drying. This operation may be carried out several times. Such operations do, however, run a risk of shattering the leaves of the hay, resulting in a loss of feed quality. Any mechanical turning, baling or other handling of dry hay must therefore be done as gently as possible.

Hay is stored drier than silage; hence changes (eg decomposition-spoilage) are less likely to occur. Hay can be stored either in the smaller, rectangular bale or the larger round bale. The crop is cut and then allowed to dry out for a period of time before baling. This drying out period is very important as wet hay is more prone to spoiling and sometimes spontaneously combusting in a storage situation. It is more difficult to make good quality hay in wetter climates.

Hydroponic fodder

Fodder crops have been successfully grown under intense cultivation using hydroponics. Hydroponic fodder may have some advantages:

- It can be produced under controlled conditions (eg inside a greenhouse) all year round, and during abnormal conditions such as drought or extreme cold
- It makes more efficient use of water supplies
- Protein content of hydroponic food may be significantly higher than the same plants grown in paddocks
- Plants may be grown in tiers (with artificial lighting), allowing much greater production per unit area
- Growth rates can be accelerated, allowing greater production per unit area, per year
- Hydroponic production can be designed to need less manpower

The main disadvantage of hydroponic fodder cropping is that the establishment costs can be relatively high. Though hydroponic fodder production is not yet widely practised, it has been used successfully for growing a variety of fodder plants including oats, wheat, rye, barley and sorghum.

If you are interested in learning more about hydroponics read *Commercial Hydroponics* by John Mason, published by Kangaroo Press.

Tobacco seedlings. Tobacco can be used as a natural pesticide.

Organic pesticide containing Neem and slow release organic fertiliser.

These vegetables are being grown as a polyculture, which is less susceptible to pest and disease outbreaks.

Crops grown in monoculture. Polycultures are less susceptible to pest invasion.

A farm dam overrun with water weeds. Some of the most notorious water weeds are *Salvinia molesta, Azolla pinnata* and *Eichornia crassipes* (water hyacinth).

Garlic flower and seed. Garlic is used as a natural pesticide, both in companion planting and by applying it in liquid form to plants.

Marigolds are useful as they reduce the impact of soil nematodes.

Lantana camara – a formidable weed and a real problem where there is light available at the edge of farm tree corridors.

Cattle on mixed pasture.

Bush tucker – an Australian example of agricultural diversification.

Sheep can be used to graze stubble before direct drilling of seed. This reduces the chance of machinery being clogged by excessive stubble.

Legumes, such as white clover, are useful for fixing nitrogen in the soil.

Field pea, for use as a cover crop.

Pinus radiata (Radiata pine) is a useful agroforestry species in temperate regions. In other areas, it has become a weed.

Oats can be used as a cover crop.

Hemp – an environmentally friendly plant crop.

Poplars and other common species can be used for windbreaks in agricultural areas.

Tubestock plants can be an inexpensive way for farmers to establish trees on their property.

Tree planting on agricultural land.

Rotational grazing should be planned so that grazing animals reach feed at its optimum growth stage. This is a rye–grass hybrid.

Emu farm, New South Wales, Australia.

Panacur 100, used for treating cattle and horses against worms. Regular drenching is an important component of integrated pest management programs. Good planning and husbandry will reduce the impact of worms with minimal use of chemical treatments.

Rotational grazing helps to reduce pests and diseases in animals. Securely fenced paddocks are a must.

Secure and safe fencing is a must for livestock production.

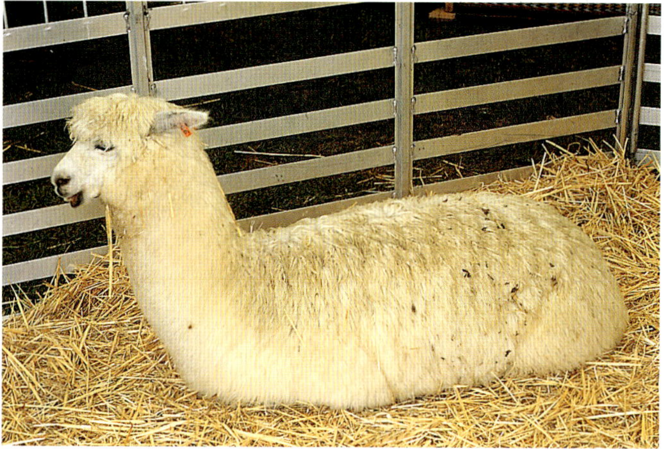

Alpacca.

Ducks on an irrigation channel. Ducks will help to control snails, slugs and insect pests.

Free range geese.

Boer goat.

Angora goats.

Giant Malaysian River prawn produced by aquaculture.

Artificial fertilisers on pasture can inhibit the uptake of some nutrient elements required by horses.

Cattle grazing on roadside verges during drought.

9

Managing plants – Tree plantings and windbreaks

Trees play an important role in ensuring that our agricultural lands remain in good health. Much of Australia's farmland has been extensively cleared. Many of our remaining trees in agricultural areas are in poor health, many suffering from die-back.

Important reasons for having trees on farms

Erosion control

Trees help control or reduce erosion in several ways, including:

- By their roots binding soil particles together
- By acting as windbreaks, decreasing the wind's ability to dislodge and move soil particles
- Acting as a physical barrier, trapping moving soil particles
- Reducing the erosive potential of rainfall by providing a protective cover over the soil below

Intercepting rainfall, which then either:

1. Evaporates back into the atmosphere without ever reaching the ground
2. Drips slowly from the tree foliage, reducing the potential for surface runoff (longer time available for water to infiltrate into the soil), hence reducing the likelihood of surface erosion
3. Flows down the branches and trunks of the trees, eventually reaching the ground with far less erosive power (energy) than if it had dripped or fallen directly onto the ground surface

Lowering watertables

Trees help lower watertables, reducing waterlogging of surface soils and salinity problems. Clear felling on farm land has resulted in the rising of watertables to the detriment of crops and other native plants. This has become a major problem for vast areas of Australia. Saline soil hinders roots development and nutrient and water uptake due to high salt content.

Sheltering stock

Trees provide vital shelter for farm animals. Stock suffering from heat stress are more likely to pollute dams and waterways. Research shows that shelter can improve milk production, ram fertility and stock liveweight. Shelter also reduces deaths of lambs and sheep from exposure during cold or wet weather.

Windbreaks

Windbreaks protect grazing animals and crops from windborne debris (eg damaging sand particles). They also help protect against cold or hot winds that can damage crops, and require stock to expend a lot of energy trying to cool or warm their bodies, rather than using the energy to produce more growth (increasing yields).

Timber

This could be in commercial plantations. Softwood timbers are commonly planted by forestry departments to keep up with building industry demands. Some farmers have been planting native timbers, not only gums, but other prized timber species that are no longer harvested from rainforests.

The concept of planting your own superannuation scheme has become popular for individuals with adequate land and suitable soil. Timber harvested on your own land may also be used for fencing and other simple structures around the property. Some people have been known to harvest their own timbers to build their family home.

See also the later section on agroforestry.

Firewood

Firewood may be grown both for on farm use and as a commercial crop. This reduces the reliance on our remnant forests. The firewood may also be smoulder-burned to supply charcoal to the nursery and other industries.

Fodder

Some tree species may provide supplementary feed for livestock, particularly during harsh times such as drought.

Honey production

Native and exotic trees can be used to produce honey. Distinctive flavours can be marketed, such as sugar gum, leatherwood, yellow box, etc.

Improving soils

Leguminous trees (eg Acacias, Casuarinas, Robinia, honey locust and Cassia) increase levels of nitrogen in soils through the action of nitrogen-fixing bacteria on their roots.

Most trees, like other deep-rooted plants, are capable of taking nutrients from deep in the soil profile and lifting them up into the leaves, which in turn fall to the ground. This, in effect, acts as a recycling system for nutrients that have been leached deep into the soil profile.

Increasing rainfall

It has been reported that treed districts receive more rainfall than nearby non-treed districts in the same area. These reports are based on large land areas, not small acreage lots. In high altitude areas the foliage canopy of tall trees may, at times, penetrate cloud layers. Moisture from the clouds may condense on the tree foliage and drip to the ground, thereby effectively increasing rainfall in the area.

Firebreaks

See later section on firebreak design.

Wildlife habitat

See later section on creating wildlife corridors.

Agroforestry

Agroforestry is the growing of trees on farms for both commercial harvesting and improved landcare outcomes. Agroforestry projects can be incorporated with sustainable farming practices such as wildlife corridors, salinity reduction and erosion control.

Agroforestry is usually a long-term project, taking up to 40 or 50 years before returns are achieved. Indeed, some farmers plant trees for harvesting by the next generation. This situation makes it difficult for many farmers to justify the time and expense required in establishing an agroforestry project. Nevertheless, the environmental benefits of timber lots can improve farm profitability during the time it takes for the trees to reach maturity.

In some instances, agroforestry plantings can be profitably harvested within a shorter time frame. For example, the vineyard industry has a high demand for pine poles of a particular diameter – too small and they are not strong enough, too large and they will prevent mechanical harvesters from operating. Pine poles suitable for vineyards can be harvested within 10–15 years after planting.

Profitability

Although some of the long-term increases in farm profitability through improved land management are hard to calculate, the economic viability of agroforestry projects should be carefully considered before fencing and planting begin.

Firstly, there must be a suitable site for growing the trees. It is difficult to identify all influencing factors, but careful analysis is essential. Some tree species (eg manna gums) can tolerate waterlogged sites, whereas some others (eg red stringybark) require well-drained sites, while still others (eg blackwood and blue gum) require fertile soils for high productivity. Topography is another factor to consider. For example, harvesting on steep slopes may be unviable and even dangerous.

Secondly, there must be a market for the timber being grown. Timber can be sold for a number of purposes. Softwood logs (eg pine) are used as sawlog timber, for plywood, for fibreboard, for veneer timber and for newsprint. Hardwoods (eg *Eucalyptus* spp.) are used for sawlogs, wood pulp and wood chips. Other timbers (eg red cedar) are in demand for high quality furniture.

Thirdly, there must be a suitable timber or pulp mill within a reasonable distance. One of the greatest expenses in agroforestry is transporting and milling the timber.

Fencing

Fencing is critical to keep out hungry stock and pests, such as rabbits and wallabies, which can destroy a crop of young trees. At least one good strong, well secured gate should be provided, so that you can get into plantings for maintenance work and to allow stock in to feed (at selected times) on the fodder trees, as well as grass, etc beneath the trees.

Design

Careful planning of an agroforestry project is vital for producing a profitable crop. The selection and mix of species and the planting design should be determined both by the economic viability of the crop and the need for sustainable outcomes. The main decision to make is whether all the timber will be removed, or removed in stages, or whether some will be retained for habitat, windbreaks, etc.

Most timber logs will have to be harvested and removed by machines, so the planting design must allow room for them. For this reason, many agroforestry projects have trees planted in easily managed straight lines.

Trees grown close together are likely to grow straight and tall to reach the light. Once this has been achieved, the trees can be thinned to allow for the proper development of the remaining plants.

In situations where large numbers of trees are needed for the treatment of salinity or waterlogging, it may be preferable to plant only a single species.

Some agroforesters choose to plant a mix of species in their timber lots. This makes management, pruning and harvesting more complicated, but may be desirable for providing ongoing shelter, honey production, etc.

It is very important to be realistic in the number of trees to be planted. Too many trees and the project will become unmanageable, and even become a potential fire hazard.

After a tree species has been chosen, seedlings (or seed) should be obtained from a supplier of quality plants. Plants from poor genetic stock, or plants that have been left in the pot too long and become 'pot bound', should be avoided. In many cases stock will have to be ordered some months before you are ready to plant.

Pruning/thinning

To produce timber that is straight and without knots, it is essential that trees are pruned up to a height of 6 m. If only some of the wood lot is to be harvested, only target trees need to be pruned. Pruning should be undertaken regularly, such as once every six or 12 months.

Thinning trees may also be necessary to allow for the uninhibited growth and easy harvest of target trees. The area to be cleared will depend upon the machines to be used for harvesting. As small trees are easier to remove, thinning should be undertaken early in the project.

Thinnings and prunings from timber lots of suitable species can be used as stock fodder. In the case of Eucalypt species that contain essential oils, there is also potential for thinnings and prunings to be collected and sold.

Harvesting

The timing of harvests will depend upon the species of tree and the use for the timber. For example, most mills will only accept pine logs that are at least 18 years old that can yield logs at least 6 m in length. Generally speaking, the larger the diameter, the more valuable the tree. The quality and dimensions of trees are less important if they are to be sold for wood chips, pulp or newsprint.

It is important that logs and the remaining trees are not damaged during harvesting and transport. For this reason, harvesting and delivery is usually done by specialist contractors with suitable knowledge and equipment.

It is not necessary (and is sometimes undesirable) to harvest the entire agroforestry crop. Those trees that are misshapen, diseased, or unsuitable in some way can be left in the ground. Although they might not have commercial use, they can still provide habitat for wildlife, erosion control and other benefits.

Source: *Agroforestry and farm forestry*, by R. Washusen and R. Reid, The Benalla Landcare Farm Forestry Group.

Timber trees

Paulownia

Paulownias are extremely fast-growing trees, used widely in China for both timber and as a livestock fodder. They grow well through temperate to subtropical climates and are being grown increasingly throughout many parts of the world, including Australia.

The Chinese recommend planting on a 5 x 5 m spacing (540 trees per hectare), then thinning to 5 x 10 m). Following this pattern, the Chinese intercrop the trees with wheat or vegetables in the early years until the trees grow too large. Wider spacings may enable permanent intercropping.

Proponents of Paulownia (usually Paulownia nurseries) will suggest the first trees can be harvested (for wood pulp, packing cases, etc) after six years. Trees may be grown to 15 years to provide much larger diameter trunks, and a more marketable product. Timber may also be used for furniture, house construction, etc. This is a new product, which may or may not be in oversupply in years to come. It is, however, a tree that offers many advantages over other trees, particularly its fast growth rate, and is well worth considering for farm forestry, erosion control, shelter belts and/or fodder.

Pine plantations

There are several pine species which are valuable forestry products. They are generally reasonably fast growing, and, once established, require little attention. In temperate parts of Australia, *Pinus radiata* is widely grown. In warmer areas, species such as *Pinus canariensis* or *Pinus caribea* may be more appropriate.

Pine needles contain toxins which tend to stop other vegetation growing under these trees. This is an advantage in that weed control is relatively easy, but a disadvantage in that it reduces the biodiversity of areas where pines are planted. It is often very difficult to establish other crops in an area that has been used for pine plantations.

Eucalyptus

There are many species of eucalypts that are valuable forest timbers, but there are others which are highly susceptible to pests or diseases, or produce less desirable timber products. Eucalypts are grown extensively for timber and paper pulp production. Their main disadvantage is the long period of time required to produce a harvestable crop from some species. For example, spotted gum (*Eucalyptus maculata*) can produce commercial timber within 20 years, while Mountain Ash (*E. regnans*) typically takes over 50 years before it can be harvested. It is wise to thoroughly investigate suitable species and markets in your locality before planting any eucalypts for forestry purposes. Your local department of forestry or similar body is the best scource of information for this subject.

Acacia

Blackwood (*Acacia melanoxylon*) and black wattle (*A. mearnsii*) have potential as timber for furniture, joinery and pulp. They are also excellent for controlling erosion. Both require adequate rainfall to be suitable for use as timber.

Fodder trees

Fodder trees are those that can be eaten by stock without any harmful side effects. They are particularly valuable as a source of feed during times of drought.

Types of trees

- Legumes have the added advantage of 'fixing' nitrogen in soil (eg tree lucerne, carob)
- Non-leguminous specimens worth considering include powton, bottle trees (*Brachychiton*), etc
- Some fodder trees have the potential to self seed and rapidly spread. In some situations this can be an advantage, saving you from re-establishing fodder plantations, in other situations it can be a problem as the fodder plants may invade remnant bushland, becoming environmental weeds (this is common with tree lucerne in some areas).

Windbreaks

Windbreaks are an important means of increasing productivity and can have many long-term benefits for the local environment. A comparison of the environmental conditions in open areas versus areas sheltered by windbreaks is shown in Table 19.

Table 19 Effect of windbreaks on the environment

Open Area	Sheltered Area
High wind velocities	Lower wind velocities
High evaporation rates	Reduced evaporation rates
Low humidity	Higher humidity
High day temperatures	Lower day temperatures
Low frost risk	Increased frost risk

Windbreak design considerations

- The wind velocity reduction effect of a windbreak is felt up to 30 times the height of the windbreak away, but is most effective between two and 20 times the height of the windbreak, on the downwind side. For example, the most protected area behind a 2 m tall windbreak would be from 4 m to 40 m downwind from the windbreak.
- If you have a living windbreak (ie plants), other plants in the immediate vicinity of the windbreak (on both sides) may have their growth reduced due to competition for light, water and nutrients from the windbreak plants. However, for plants at the equivalent of at least twice the height of the windbreak in distance downwind from the windbreak, growth rates are increased.
- A windbreak consisting of dense foliaged plants, or a solid timber or brick wall will deflect wind directly backwards, as well as up and over, creating strong turbulence both in front and behind the windbreak. In comparison, a more permeable windbreak, such as more open foliaged plants, slatted fences or windbreaks made from shadecloth, allows the smooth flow of air through, and up over and past the windbreak with little turbulence.
- Be careful not to position your plants too close together or they will self prune (drop branches) and become too open to be effective. This is particularly important for some conifers, which will self prune their lower branches if planted too closely together, creating a gap below the plants, which can allow the wind to be channelled through, actually increasing the problem of strong winds. This is a common occurrence in many farm windbreaks.
- Don't cut plants back too heavily. It is better to do frequent light trimmings, otherwise they will become too open foliaged to be effective. Larger windbreaks comprised of plants also have advantages and disadvantages with regards to fire risk. If plants are highly flammable (often have high levels of volatile oils) they pose a real risk to buildings and structures downwind. If they are comprised of fire resistant plants (generally having a high moisture content) and positioned to deflect winds they can provide significant protection downwind, including protection from radiated heat.
- At the end of a windbreak, or where gaps occur (eg driveways) wind velocity may actually be increased, creating a wind tunnelling effect.
- Increasing the width of a windbreak may decrease its effectiveness by reducing its permeability.

For a windbreak to be most effective:

- It should be as long as possible, with no gaps or breaks. The area protected by a windbreak varies with the square of the unbroken length of the windbreak. So, for example, a windbreak of 20 m will protect an area about four times as large as a 10 m windbreak, while a 40 m windbreak will protect an area about 16 times that protected by a 10 m one.
- It should be as long and as high as possible (to protect a bigger area).
- It should have no gaps at the bottom (otherwise wind will funnel under the windbreak, increasing in velocity and increasing the likelihood of wind damage).
- It should be permeable to air flow (it allows the passage of some air but reduces its velocity).

- It should have a cross-section (the profile of the windbreak when viewed from its end) that results in the smooth passage of some wind up and over the windbreak (known as an aerofoil cross section). To achieve this you would have low growing plants at the front of the windbreak, slightly bigger growing plants behind them, and taller plants at the back.

Windbreak plants

Many native plants are suitable for windbreak plantings. Lists of suitable species for your area are usually readily available from local branches of your relevant government departments such as agriculture, forestry, primary industries or conservation. In addition, groups such as Greening Australia and local landcare groups usually have similar lists. A few of the more commonly used species are listed below.

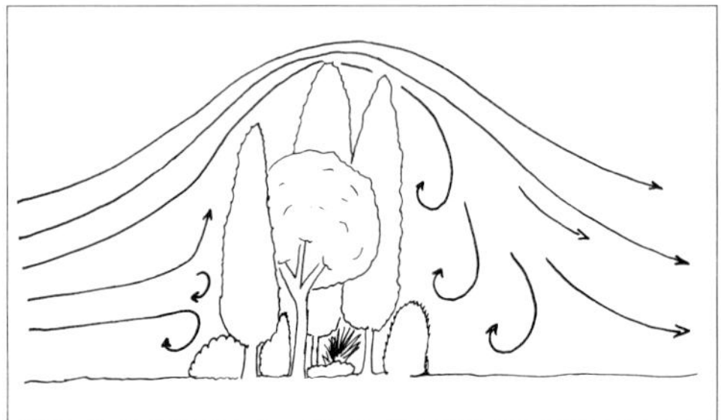

Figure 9.1 A dense windbreak

Figure 9.2 A permeable windbreak

Windbreak natives for temperate Australia

Acacia floribunda, longifolia, mearnsii, pravissima, saligna
Acmena smithii
Agonis flexuosa
Allocasuarina (most varieties)
Callistemon citrinus, pallidus, salignus, viminalis
Callitris (most species)
Correa alba
Eucalyptus alpina, camaldulensis, forrestiana, globulus compacta, leucoxylon, pauciflora, polyanthemos, viminalis, viridis
Grevillea rosmarinifolia
Kunzea ambigua, flavescens
Leptospermum laevigatum, lanigerum, scoparium
Melaleuca armillaris, elliptica, ericifolia, huegelii, hypericifolia, nesophila, squarrosa, stypheloides
Myoporum insulare
Pittosporum undulatum
Westringia fruiticosa

Natives suitable for windbreaks in tropical and subtropical climates

Acacia (various species)
Acmena smithii

Allocasuarina (most species)
Angophora costata
Araucaria cunninghamii, heterophylla
Archontophoenix alexandrae
Backhousia citriodora, myrtifolia
Banksia integrifolia
Brachychiton acerifolius, populus
Buckinghamia celsissima
Callistemon formosus, viminalis
Callitris collumellaris
Castenospermum australe
Cupaniopsis anacardioides
Eucalyptus tereticornis, tessellaris
Flindersia spp.
Grevillea robusta
Harpullia pendula
Hibiscus tiliaceus
Leptospermum flavescens
Melaleuca leucadendron, linariifolia
Melia azaderach
Pittosporum revolutum, rhombifolium, undulatum
Pleiogynuim timoriense
Podocarpus elatus
Syzygium (many species)
Vitex ovata

Firebreaks

Fire-prone areas

Much of Australia is subject to hot, often very dry, summers with strong gusty winds. In addition, much of the Australian flora has a high level of volatile oils in its foliage. These factors combine to make parts of Australia amongst the most fire-prone areas in the world.

To ensure maximum safety for your property you would need to remove all burnable material for a considerable distance away from whatever you are trying to protect. This could result in a barren, unattractive landscape that most property owners would not consider desirable. By careful selection and placement, however, it is possible to have plants nearby while still maintaining an acceptable safety level.

How to arrange plants

Careful placement of plants can significantly reduce the impact of fire. The immediate area around buildings should be free of trees and other combustible materials. Lush, well-watered lawns, paved areas, driveways, etc. in this area can provide an effective barrier to the passage of fire.

A fire-retarding shelter belt placed at right angles to prevailing winds will also protect buildings (do not place the shelter belt too close to buildings – a minimum distance is the height of the shelter belt, although ideally the distance should be three to five times the height of the shelter belt). The shelter belt will act to reduce the wind which fans the fire, will deflect heat and smoke and will catch burning airborne material. The shelter belt should be made up of fire-tolerant or resistant species.

Those people planting shelter belts or corridor plantings to provide habitat, safe passage, food, etc for wildlife, may have to compromise a little in their design (see 'Points to Remember'). Wildlife corridors may have to be sited sufficiently far way from your buildings and structures so as not to pose a fire risk.

Distances from buildings

Keep trees at least the same distance as the height of the mature tree from any buildings, eg if the height of a tree is 20 m when fully grown, then it should be planted at least 20 m away from any building (if the tree falls, then burning branches won't hit the building).

Prevailing winds

The prevailing winds will affect the way fires will travel and where ash and burning embers fall. It is important to note that prevailing winds may vary from season to season, and place to place, although days of extreme fire danger are usually characterised in south-east Australia (a major area for bush fires) by hot gusty northerly or north-westerly winds with southerly wind shifts later in the day.

Vehicular access

Access routes to dams, pumps, roads, etc should be kept free of trees and flammable material. This includes all routes of escape. Areas around pumps should also be kept free of flammable material.

Fire resistant plants

The following types of plants are less likely to catch alight and burn in a bushfire:

Plants with high salt content (eg Tamarix, Rhagodia, Atriplex, *Eucalyptus occidentalis*, *E. sargentii*)
Plants with fleshy or watery leaves (eg cacti)
Plants with thick insulating bark
Plants that have their lowest branches clear of the ground
Plants with dense crowns

Fire prone plants

Plants which are more likely to burn include:

Those with fibrous, loose bark (eg stringybark eucalypts)
Those with volatile oils in their leaves (eg most eucalypts, callistemons, melaleucas)
Those with volatile, resinous foliage (eg many conifers)
Those with dry foliage
Those which retain or accumulate dead leaves and twigs

Maintenance of firebreak species

For firebreaks to be effective, they need to be well maintained. Remember to:

- Water trees in summer (this helps keep moisture levels in the plant high)

- Fertilise your plants regularly in summer if soil is moist or rainfall is adequate; a plant that has lush green growth is less likely to burn
- Have a hose ready at all times and ensure water is readily available
- Only use mulches near buildings that will not burn readily. You should remove twigs, leaf litter, etc from the ground. A compact mulch of stone, or even woodshavings, is not generally a problem, but leaves and twigs can be. Leaf litter can be dug in or composted to prevent it burning.
- Remove flaky loose bark from trees; smooth-barked trees are less likely to catch fire
- Prune lower branches so that burning debris under plants can't ignite foliage
- Remove dead trees and fallen branches
- Prune off hollow limbs or fill cavities (hollow trunks, depressions where branches break and rot gets in) with expansion foam or concrete – or remove the plant ... fire can catch in such hollows and the tree may smoulder for some time without you knowing it.
- Have succulent groundcover, lawn or gravel under large trees or regularly slash or cut any underlying scrub and grass prior to the fire season to remove potential fuel for fires.

Habitat corridors for wildlife

Why create a wildlife corridor?

Large areas of indigenous vegetation have been cleared for housing, agriculture, industry, and other uses, hence there is greatly reduced habitat left for native wildlife. Many of these native vegetation fragments are often small and isolated from one another by barriers such as open pasture, housing, roads, and waterbodies (eg dams). These are sometimes known as 'island' habitats.

It should be noted that wildlife is more than just birds and mammals. It also includes insects, reptiles, spiders and micro fauna such as earthworms. Without this diversity of smaller animals, many larger animals will not be able to survive.

Wildlife constantly move:

- Looking for seasonally available and new sources of food
- Looking for shelter/protection
- Searching for mates
- Dispersal of young to new ranges

In island habitats, there may be no adjacent habitat to forage in, or for animals to roam and disperse. The vegetation fragments may not provide all the resources an animal species requires (eg food, water, shelter/protection and breeding). Island plant communities are also vulnerable to catastrophic events, such as pests, diseases, clearing, bushfires, and to gradual changes, such as inbreeding or climatic variation.

Habitat corridors provide links between these isolated island communities. This allows migration to replenish a declining wildlife population (increasing numbers giving better chance for some to survive and reduce inbreeding) and also allows recolonisation where a species may have become locally extinct (extend the local range).

Other benefits of wildlife corridors

There are not only benefits for indigenous vegetation and wildlife, but also considerable benefits to local land owners. Creating such corridors can also:

- Help reduce erosion (eg in gullies, stream banks, and on exposed ridges)
- Help reduce salinity problems
- Reduce nutrient runoff into streams
- Provide windbreaks or shelter belts for stock and crops:
- Greatly improve yields due to reduced heat or cold stress of stock
- Reduce physical damage to plants (eg young seedlings, flowers on fruiting plants) by wind (direct wind effects and sandblasting effects)
- Increase birth rates of stock (up to 50% increases recorded in lambing rates in some areas)
- Provide timber and firewood
- Help improve water quality
- Help mitigate floods
- Improve recreational fishing

Where to establish wildlife corridors

- Corridors may exist anywhere between habitat islands of any size, even as small as a few old remnant eucalypt trees that may provide valuable hollows, or linking smaller patches to eg a larger state forest.
- Remnant wetland environments (eg marshes, swamps, lakes) can also be linked with other vegetation corridors, providing improved access for wildlife to important water sources.
- They are best designed, where possible, to follow natural contours (eg rivers, ridges)
- They might incorporate other farm plantings (eg windbreaks, timber lots)

Types of wildlife corridors

- Natural corridors that follow land contours (eg ridges, streams, gullies)
- Remnant vegetation such as those along roadsides, railway reserves and disused stock routes; these corridors often follow straight lines
- Planted corridors include such things as farm shelter belts and windbreaks; these are generally created for purposes other than creating wildlife habitat, but this can be incorporated into the design through careful selection of plant species

Wildlife corridor design

- Preserve or restore natural corridors (eg gully lines, stream banks); streamsides are high value areas for wildlife. Limit stock access to riverbanks to prevent erosion and allow for regeneration of riverside vegetation.
- Wherever possible build onto or restore existing corridors as they will have existing populations of local flora and fauna, increasing the rate of species spread.
- The wider the corridor the better (eg at least 30–100 m wide) (see section on 'edge' effects).
- Corridors are more effective when they link up with larger habitats with few or no gaps (eg roads cutting through).

- Use local (indigenous) plants. These are adapted to local conditions (eg soil, climate, fire regimes) and local fauna are adapted to them. This also preserves the biodiversity of local flora. Indigenous plants generally have low establishment costs in comparison to introduced species and have minimal weed potential.
- Incorporate all forms of vegetation (eg shrubs, grasses, rushes, groundcovers, climbers), not just trees. In some grassy forests of northeast Victoria, for example, there may be four species of Eucalypts and between 70 and 100 understorey species. This means that the understorey represents over 90% of the biodiversity of the vegetation in this ecosystem.
- A network of corridors is more effective than single links. They increase opportunities for migrations and reduce the risk of links being broken (eg bushfires, subdivision and subsequent clearing of some blocks).
- Fencing to restrict grazing of corridor vegetation by domestic stock is very important, but be careful not to restrict movement of wildlife.
- Consider habitat (eg rocks, hollow logs, leaf litter) for animals that may be slow in migrating (eg small ground dwellers such as lizards and snakes). Consider the provision of artificial nest boxes, or placement of hollow logs within new plantings.
- Cooperative action between local landowners may be necessary. Such cooperative efforts can make the best use of available resources, and allow for the most effective links between remnant patches.
- Agroforestry projects can be be positioned to link remnant vegetation patches, and also to act as a buffer around larger remnant vegetation patches.

Edge effects

'Edge effect' is a term used to describe what occurs with regard to vegetation and wildlife when one type of vegetation shares a border with another. They may occur naturally (eg forest grading into woodland, or streamside vegetation to drier nearby slopes, and burnt and unburnt areas); or they can be man-made, such as pasture abutting forest, or roads through forest. Some edge effects can be positive in terms of native flora and fauna, but most tend to have negative effects. Edge effects are most likely to have an influence on narrow strips or small remnant areas. In terms of corridor plantings, the wider the corridor the less the impact of 'edge effects'.

What can happen at edges?

- Solar radiation, air and soil temperature, wind speed and humidity levels can all be altered, leading to stresses on existing vegetation and change in the types of plant seeds germinating.
- As vegetation patterns change near edges, so usually do the types of wildlife that inhabit those areas. Edges can be important for some species, providing shelter, nest sites, perching and observation points (eg parrots feeding on grass and grain seed; eagles on rabbits; and kangaroos and wallabies on grasses).
- Species with wider tolerances take over near edges while less tolerant species only survive in 'core' areas away from edges. In narrow corridors or small remnant patches these core species are generally absent.
- Aggressive edge-dwelling species such as Noisy and Bell Miners may invade and displace former inhabitants.

- Pest animals such as foxes, cats and dogs tend to move along and hide out near roads, tracks and cleared areas.
- Invasive (weed) plants can readily move into remnant vegetation and corridor plantings from adjacent agricultural, industrial or residential areas.
- Chemicals and fertilisers can drift from agricultural areas into edge areas.
- Erosion and altered water runoff characteristics (eg drains) can damage and undermine the soil in edge areas.
- Stock can trample and graze edge areas.
- Litter (eg from roadsides) can pollute habitat areas.
- Noise and movement from traffic and agricultural activities can disturb animals that require quiet to breed and feed.

In general:

- The longer the edge, the larger the area that is vulnerable to disturbance.
- The more angular the edges the greater the edge effect; corners increase disturbance; rounded corners and regular shapes minimise edge effects.
- The smaller the area, the greater the risk of impacts occurring throughout the vegetation, with the 'core' habitat being destroyed.

Source: *Grow your own wildlife*, by P. Johnston and A. Don, Greening Australia Ltd.

Tree planting methods

There are two main ways to plant trees on farms: by planting seedlings or by direct seeding. Seedlings are usually bought in 'tubes' of approximately 4–5 cm diameter. Use of larger plants is generally uneconomic.

Preparing the site

Site preparation has a major influence on the success of any planting project. Good site preparation will include:

- Fencing – all new plantings are vulnerable to grazing by stock, pests and native animals. To minimise the problem, either the entire area or the individual plants should be protected. Fences must be strong and secure (to keep out rabbits, fences will need to be buried underneath the soil). Tree guards have also been successfully used to stop animals biting off leaves before the plant has become established.
- Weed control – Competition from weeds can seriously undermine a planting project. Where possible, weeds should be removed from the planting area before work commences. This is usually done either by cultivation or by the application of herbicides (see Chapter 5).
- Ripping – In hard or compacted soils it may be necessary to 'rip' the soil. Ripping is the process of digging rows in the paddock to improve root penetration in the soil. Cross ripping will reduce the likelihood of roots growing along the rip lines and making the trees vulnerable to being blown over in strong winds.

Planting seedlings

Planting seedlings is a well-established and proven practice. It is more expensive than direct seeding, but it is better suited to projects where plant spacing is important, such as agro-forestry.

Holes for seedlings are often dug by hand but specialist tools are also available. One of these is the 'Hamilton planter', which removes a tube of soil, into which the seedling is then placed. Named after the town in Western Victoria where it was invented, this device is only suitable for deep fertile soils.

There are also machines that have been designed for planting seedlings. A cultivator digs a furrow in the soil, while a person on the back of the vehicle steadily places seedlings in a chute directly above the furrow. The plants then fall directly into their planting holes.

In some situations it will necessary to water the freshly planted seedlings. However, where possible, planting should be undertaken when there is sufficient soil moisture.

Tree guards are often necessary to protect young seedlings.

Direct seeding

Direct seeding is becoming increasingly popular as a method for establishing windbreaks and habitat corridors. It is comparatively cheap and will produce a random distribution of plants, giving a more natural appearance than plants in rows. Plants grown from seed are also less likely to be blown over in the wind.

One of the great advantages of direct seeding is that it can be done using conventional farming equipment.

The disadvantages of direct seeding include the difficulties the plants can experience from competition by weeds, as well as grazing by insects and other animals. Direct seeding can also result in an uneven distribution of plants. It requires more seed than planting and is limited to those species that germinate easily from seed. A further disadvantage is that it is only practical in areas with suitable rainfall.

Direct seeding can be undertaken in belts (eg for windbreaks), in rows and in patches. (To minimise the edge effect these patches should be circular.) After the weeds have been controlled and the area cultivated, seed is distributed either by hand or with a machine such as a fertiliser spreader. The site is usually then harrowed or rolled to lightly bury the seed.

Direct seeding is usually undertaken during suitable conditions:

- In winter rainfall zones, areas with high rainfall can be sown between August and October. In lower rainfall areas, sowing should be undertaken between May and August. At this time frosts can be a problem for some plant species.
- In summer rainfall zones, seeding is typically undertaken before the wet season. This avoids cultivating wet soil and allows plants to become established before the dry season. However, there is still some uncertainty about the best time to sow in subtropical districts. In southern Queensland, direct seeding is often undertaken in March to April, during the middle of the wet season.
- Some seeds require specific temperatures or light conditions (short or long days) for germination.

Source: *Direct seeding of trees and shrubs*, by G. Dalton, Primary Industries (SA).

10

Managing animals in a more sustainable way

Sustainability aims to optimise the long- and short-term productivity of a property, whereas traditional farming has often aimed to optimise the short-term production of individual farm products.

Possible problems of livestock production

- Degradation of pasture through trampling, introducing weeds (eg in feeds)
- Destruction of habitats
- Competition for forage material between domesticated and wild animals
- Loss of biodiversity
- Pests eg foxes, rabbits, escaped domestic stock (camels, pigs, horses, goats, cats, dogs, etc)

To sustain livestock production on a property, the following must be done:

1. Select breeds of livestock appropriate to the site
2. Control overstocking
3. Use an appropriate production system
4. Apply appropriate landcare practices to sustain the condition of the land (eg subdivision fencing according to soil types and land use)

Breed selection

Before selecting a breed, determine the type and quantity of feed and water available. Discuss any proposed selection with people who know the local area. Consider the way in which the livestock might need to be managed (eg fencing requirements, frequency of moving animals). You need to have the manpower, equipment and financial resources to manage the chosen breed in a sustainable way. If you don't have adequate resources, you might be better to choose a different type of animal (eg goats are good in a paddock for a while to eradicate weeds but, at a certain stage, they can start to cause degradation of land).

Stocking rates

This refers to the number of animals that can be supported by a specified area (ie head per hectare). The optimum stocking rate of a property may vary from month to month and year to year according to seasonal changes and unproductive periods such as drought.

Supplementary feeding and watering may allow stocking rates to be increased on a property or at least maintained during periods of poor pasture growth. Animals may also be put elsewhere under agistment at times to relieve their influence on the property.

Problems can develop if animals are allowed total freedom on a property. For example, they may congregate in one particular area, causing erosion, or they might only eat one particular pasture species, causing a change in the pasture composition. Generally, animals are restricted to different areas at different times.

Fencing

Fencing is necessary to contain stock but it can also injure them. Fencing is traditionally six strands of barbed wire, however most vets oppose using barbed wire because it injures stock, affects their health and damages hides (making them less valuable for the hide industry). One or two strands of barbed wire will generally be adequate, with the remaining strands being plain wire.

The best fence is post and rail (four rails for small animals and two or three for large animals). In most instances, post and rail fencing is too expensive to be economically viable, except for intensive areas such as stock yards or feedlots. Post and rail may also be used to contain particularly strong or valuable animals, such as stud horses or bulls.

Electric fences are relatively inexpensive and increasingly popular. They can be moved with relative ease, providing much greater flexibility and allowing paddocks to be reconfigured frequently. Electric fences can, however, be a fire risk if the fencelines are not routinely checked.

All types of fencing require maintenance. Fences do move and need straightening, while gates deteriorate and need to be repaired and reswung. Electric fences, perhaps, take less effort to maintain than others.

Production systems

There are no hard and fast rules about what production system is most sustainable for a particular type of animal. The following examples provide an insight into systems which have been used in the past; however the system you choose for your property is better tailor-made to suit the conditions there.

Rotating uses of a paddock

This involves using a paddock for different purposes at different times; for example, growing a cash crop in one season; grazing in the next, followed by growing a green manure cover crop before planting another cash crop.

Multiple use of a paddock

Paddocks can be used for two or more different purposes at the same time; for example, grazing under a tree crop, or intercropping annual crops between permanent plantings such as fruit trees or vines.

Low intensity stocking

This involves keeping stock numbers at a level to be sustained by the poorest seasonal conditions. It works well on large properties where land is cheap but may not be financially viable elsewhere.

Free range

This involves allowing animals to run free on a property, or part of a property. Productivity levels might not be as high, but it is a low input system, usually with significant cost savings on manpower, equipment and buildings. Predators can be a problem, particularly with smaller animals, and animals may be more difficult to handle because they are handled less. Pigs and poultry farmed under a free range system are often healthier, less susceptible to passing diseases from one animal to the next, and able to exercise better than in intensive systems.

Intensive confinement

This involves keeping animals in a confined area (eg horses in stables, poultry in sheds or cages, pigs in sties, beef and dairy in feed lots). This system requires high inputs. With animals living close together, diseases can spread fast, so chemical controls are frequently used, water and feed need to be brought to the animals, wastes need to be removed and disposed of, areas need to be cleaned and perhaps sterilised, and animals may need to be exercised periodically. Intensive systems do not use as much land, but they use more of just about everything else, and have a greater potential to develop problems such as epidemics or land degradation.

Integrated farming-grazing crop residues

Grow a crop such as corn or wheat, harvest the crop, then bring animals onto the paddock to graze on the crop residue. Concern is sometimes expressed that grazing between crops may result in excessive use of the land resource, resulting in degradation effects such as soil compaction, reduced soil organic content, and reduced crop productivity.

In the 1990s, the University of Nebraska conducted studies into these concerns which showed no decrease in crop production; however there was an increase in soil compaction and a decrease in percent residue cover. It appears that the effect of grazing on crop residues is minimal (if anything) over three or four years, provided the ground is not excessively wet. Problems become exaggerated if ground is wet. Residue grazing is usually done with lighter, less disruptive animals, such as weaners and yearlings.

Combinations

In cold winter climates, animals are confined over winter to protect them from extreme cold, then let out to graze in spring. Pigs or beef raised in paddocks are sometimes confined for a short period prior to slaughter to 'finish' and improve the final meat product.

Landcare practices

By developing an environment that better suits the livestock being grown, you are able to maintain and perhaps even increase production.

- Animals become less stressed if provided with protection from extreme weather conditions (eg through planting shade trees and shelter belts).
- Diseases are less likely to develop if animals are isolated from waste products (urine and faeces); this is particularly important in areas of more intensive production (ie. where stocking rates are high) or where animals congregate (eg dairies, feeding and watering troughs, poultry nests, stables or other shelters).
- Vulnerable areas may need to be fenced. Steep slopes are more sensitive to erosion. Sources of drinking water are more sensitive to contamination from livestock.
- It is important to maintain as great a diversity of animal and plant populations as possible, both in the wild and on farms. Varieties which were not valuable in the past have proven a valuable source of genetic material from which to breed many modern farm animals and plants. Similarly, varieties which might not seem important today will probably be extremely important to breeders of the future.

Pastures

In the past, pastures were managed with the aim of achieving maximum plant growth and productivity. This is increasingly changing. The primary aim today should be to care for the plants and the soil; and in turn the grazing animals needs are better met by improved soils and plant productivity. Sustainable pasture management depends upon a good understanding of the biological processes involved in pasture growth and health. Ideally it involves:

1 Observing the factors that affect the condition of a pasture, ie soil, plants, animals, weather patterns.
2 Influencing the factors that affect the condition of a pasture to curtail any degradation. For example, if soil condition is deteriorating, fertiliser may need to be added, or the number of animals grazing may need to be reduced. If growing conditions for plants are becoming strained, the number of animals per hectare should be reduced.

Sward dynamics is the study of growth responses to different grazing management practices. Most sward dynamic research has been confined to common, cool temperate pastures (eg perennial ryegrass and/or white clover) growing in relatively uniform climates such as England or New Zealand. There is only limited information available about how pastures respond to different treatments in warm climates, unpredictable climates, or with less common grasses (eg native pastures in Australia). Given the lack of solid information, sustainable management in many pastures may require close observation coupled with a degree of caution.

Sustainable pasture varieties

Understanding and managing pasture is highly complex. Pastures differ in terms of both the mix of plants of which they are composed and in the way those plants grow (ie general plant health and vigour). For effective grazing, the pasture needs to match its use: the number and type of animals being grazed should be appropriate.

Some of the more important pasture varieties in Australia are:

- Grasses: ryegrass, fescue, cocksfoot, kikuyu, paspalum, Phalaris, prairie grass, sorghum, oats, buffel grass
- Legumes: lucerne, clovers, vetches, lotus, sainfoin
- Other plants: saltbush, chicory
- See section on cover crops (Chapter 9) for more information on some of these pasture varieties.

Saltbush

Saltbush is particularly useful in salt-affected, arid or semi arid areas. It is a very nutritious fodder plant. Compared with seaweed (often used as stock feed supplement), saltbush is higher in most nutrients, including iodine.

Its value as a stock feed is considered comparable to, or better than, most other feeds including clover pasture, green or dry grass pasture, barley, oats, silage or lucerne. With the exception of lucerne, saltbush is around 28% above most other pasture species in dry matter.

Saltbush also appears to have some health benefits for livestock. Sheep grazed on saltbush appear to have fewer health problems. It appears high in sulphur (a characteristic shared by garlic). This may suppress both fungal and pest complaints within the body.

At Narromine in New South Wales, a property is planted with saltbush (on 2 m x 4 m spacing) then later seeded with a pasture mix of lucerne, snail medic and *Bambatsi makarikari* grass. This treatment has been shown capable of increasing the stock carrying capacity fourfold, largely because saltbush can survive much better during dry periods.

At Donald in Victoria, a property planted with saltbush is used for grazing goats. Plants are on a 1 m x 1.25 m spacing, planted in late autumn. Here, goats are let onto saltbush for one to two hours then removed to another paddock for the remainder of the day. Even when plants are eaten back very hard, they still recover, irrespective of whether it rains or not!

The main varieties used in farm situations are:

1. *Atriplex nummularia* – old man saltbush
2. *Atriplex vesicaria* – bladder saltbush
3. *Atriplex semibaccata* – creeping saltbush
4. *Rhagodia hastata*
5. *Rhagodia linifolia*
6. *Chenopodium triagulare*

Atriplex is the saltbush of the black soil country and forms the most important genus. Rhagodia species have succulent fruits. Chenopodium favour the sandy and light-red soils.

Source: *Farm management*, by John Mason, Kangaroo Press. This book has an excellent chapter on pastures.

How much grazing?

Overgrazing can be a serious problem resulting in:

- erosion
- an increase in weeds
- a change in relative proportions of pasture species

You must watch the animals on a pasture and move them or provide supplementary feed before overgrazing occurs. Pastures will, however, respond to heavy grazing followed by a period of rest. It is like pruning a rose or fruit tree, removing part of a plant will promote a flush of new growth.

Management of grazing requires an understanding of the following terms:

- Pasture mass – measured as kilograms of dry matter per hectare (kg DM/ha)
- Occupation period – when different groups of animals are using the same paddock, this refers to the combined length of time that the paddock is grazed per rotation
- Period of stay – length of time a group of animals is left in a paddock, per rotation
- Recovery period – the period that the pasture is left without being grazed – commonly 12–50 days in reasonably fertile, well watered, temperate climate pastures

How long to graze?

The big question is: when should animals be grazed on a particular paddock, and for how long? Various formulae have been devised to calculate answers to this problem. Stocking rates are commonly stated in terms of 'Dry Sheep Equivalents' (DSE). This is affected by many factors, including rainfall, recovery periods (in turn affected by type of pasture species) and soil fertility.

In South Australia, R.J. French developed a system for determining DSE based on rainfall. An example of one way this system might be applied gives a potential stocking rate of 1.3 DSE per hectare for each 25 mm of annual rainfall which exceeds 250 mm.

Therefore:

$$\text{Potential stocking rate} = \frac{\text{Annual rainfall in mm} - 250\text{ mm}}{25 \times 1.3}$$

Source: 'Future productivity on our farmlands', by R.J. French, in *Proceedings of Fourth Australian Agronomy Conference*, Latrobe University, 1987.

Principles to follow when resting a paddock:

- Paddocks growing nitrogen-fixing plants (eg legumes) should be fallowed with nitrogen users such as grasses

- Grow a weed-suppressing crop in a paddock which just finished with a relatively non weed competitive crop
- Alternate cool and warm season growing plants in a paddock
- Allow an adequate period between repeat plantings of the same type of crop in a paddock so pest and disease problems can die out (for most crops, three years is adequate; for some, longer may be preferred)
- Grow shallow-rooted plants in an area to follow deep-rooted plants
- Alternate higher and lower users of water
- Follow heavy feeders with light feeders
- Use weed-suppressing plants (eg sorghum and oats) periodically where possible
- Use pest/disease-suppressing crops periodically where possible and appropriate (eg garlic for fungal diseases; marigolds for nematodes)
- To maintain the vigour of native pasture species, follow the general rule of 'graze half and leave half' (ie allow no more than 50% of the leaves to be removed, then move stock elsewhere)

Grazing methods

The two main methods of grazing management are continuous or rotational.

Continuous management

Here animals are left in the same paddock throughout the entire growing season (they may be moved elsewhere over winter, or when it is time to sell).

- This works well in areas with dependable climate (eg England)
- It is not appropriate when growing conditions are variable (eg areas that have spurts of growth, or dry and wet periods)
- It is not suitable in pastures containing a variety of species with different growth rates – some species can be overgrazed, and others undergrazed

Rotational grazing

Research has shown that overgrazing is related more to the time animals spend in a paddock than to the number of animals in the paddock.

Rotational grazing is usually preferable for farm sustainability.

- It has sometimes involved rotating a herd between several paddocks, ignoring the status of pasture in each paddock, sometimes undergrazing, sometimes overgrazing
- This method should be used to minimise overgrazing and undergrazing

Voisin grazing method

Voisin was an academic and scientist in France who devised a grazing method that interferes minimally with the pasture environment.

The concept involves dividing a pasture into small paddocks and rotating animals through them. The rate of rotation is dependant upon growth rate of pasture plants, and the pasture mass.

The aim is to keep plants as close to the peak of their growth curve as possible. Pre and post grazing pasture masses are estimated for each paddock, and this then forms a basis for deciding when to move animals on to the next paddock.

Reference: Chapter 8 in *Sustainable agriculture in temperate climates*, by Francis *et al.*, Wiley.

Strip grazing

Problem When animals are given a greater choice for grazing they can become selective; hence certain plants in the pasture can be eaten out and disappear, while other less-favoured species remain relatively untouched.

Answer Restrict animals to a small area and they become less selective about what they eat, so the area can be grazed more evenly.

Other areas for grazing

During drought or other difficult periods, extra temporary grazing may be found on public land or other sites in your locality.

Roadside grazing

Roadsides often provide extensive areas of suitable foodstuffs for stock. In areas where remnant vegetation is scarce there may be long stretches of grass suitable for cutting as hay, or for grazing stock on. Grazing such areas has the added advantage of reducing fire risk by reducing fuel loads. It may have a detrimental effect on any small patches of remnant vegetation that may still exist, particularly if the vegetation is very palatable for your stock.

- Control of stock is critical

  They should not impede traffic or create a safety risk (eg risk of car accidents). They can be controlled by temporary electric fencing, but this should be regularly checked, or by being herded at all times (eg by stockmen and dogs).

- It is important to check local regulations regarding roadside grazing. You may require a permit, or it may not be allowed at all. Consider the damage that may be done to the public or vehicles if an accident should occur.
- In some states, established droving routes have been gazetted in state legislation. Regulations governing grazing on these reserves should be checked out before contemplating such grazing.
- As with stock on your property, it is very important that stock grazing on roadsides have access to a suitable water supply. Access to such supplies should be controlled to prevent damage to the supply and its surrounds.

Public land

In some areas it may be possible to obtain permits or leases to graze public land. These may be temporary (eg as a means of reducing fire risk) or ongoing. Check with your local department of agriculture or land management to see if this is possible in your area.

Commercial timber plantations

Large pine and hardwood plantations may have areas of grass (eg between rows) large enough for grazing. It may be possible to obtain permission from plantation managers/owners to graze these areas. (This practice helps them by reducing fire risk and competition for their trees.)

Council land
Council approval is needed.

Industrial areas *(***around factories***)*
Consider any approval needed from local council or factory owners. Some factories allow agistment on the surrounding fields as this reduces labour and costs for slashing. Caution is needed to ensure that no toxic waste which may harm animals is dispersed onto the land.

Guidelines for raising different livestock

If you select a breed which is appropriate to your site conditions, sustainable farming can be made much easier. The following guidelines are generally applicable to most types of animals, and most breeds:

- Minimise stress to animals and health/production/disease resistance improves.
- Leaving animals in the same paddock for a long period is likely to result in higher levels of parasites (eg worms).
- All animals need some shelter, whether trees, bush or buildings.
- Moving animals periodically between paddocks will help their health and reduce the likelihood of land degradation.

For certified organic produce, some practices such as vaccination and artificial lighting are restricted. Other practices such as use of certain chemicals, using certain feed additives (eg urea) or intensive confinement of animals may be prohibited. If you plan organic farming, check the official requirements first.

Figure 10.1 Llama and deer.

Alpacas

These animals are more efficient at digesting food, less susceptible to diseases, and less likely to damage the ground than hoofed animals. Whereas hoofed animals (eg horses, cattle, sheep) cut the surface of the soil and exert a concentrated pressure on the soil surface, an alpaca has a soft padded foot that causes relatively little damage. A horse's foot exerts a force around three times that of a man's foot. An alpaca's foot exerts less than half the force of a man's foot.

Breeds

Alpacas belong to a group of South American animals collectively known as llamas.

Others in the group include the true 'llamas', 'guanacos' and 'vicunas'.

The alpaca is farmed mainly for fibre, whereas the llama, guanaco and hybrids may be grown for a wide variety of purposes including, fibre, as pets and as pack animals.

Alpacas are also used to guard other stock. Males over two years of age are particularly useful to chase foxes or dogs, protecting other livestock such as sheep.

Alpaca fleece is in many ways superior to sheep's wool. It is very clean, high-yielding, super soft and strong; it has a high lustre, comes in a range of colours, and is less likely to require chemical or dye treatments than wool.

Husbandry

Temperament problems can be avoided by proper training. (Halter training is advisable.) Vaccinations and parasite control are usually advisable.

Feeding

An alpaca can digest feed with twice the efficiency of beef cattle and 40% more efficiently than a sheep.

Problems

Conformation problems can occur, such as cloudy eyes, unaligned upper and lower jaws, bent legs, and kinks or lumps on the body.

Heat stress can be a major problem with alpacas, particularly in warmer climates such as Western Australia and Queensland. Signs include actively seeking shade and water, panting and nasal flaring. Animals with a shorter fleece are better able to cope with heat than those with 8 cm (3 inches) or more fleece.

Johne's disease, a chronic diarrhoeal condition, can sometimes be a problem.

Advantages

The alpaca is free from footrot and flystrike, requires no crutching, no special fencing requirements or shedding, and has trouble-free birthing. They are known to be hardy and disease resistant and protective of their young from dogs and foxes.

Unlike sheep, the alpaca has a naturally bare anus and vulva, hence it doesn't require routine treatment for flystrike. Tails do not need to be docked, because they are naturally short. Alpacas also appear more resistant to internal parasites than traditional livestock.

Alpacas exhibit a natural aggression toward foxes, both in their native South America and in captivity as livestock, and will chase them from a paddock. (Some farmers even put alpacas in a paddock with sheep in the hope that they will chase foxes away during lambing.)

Aquaculture (freshwater)

It is sometimes possible to diversify farm income by utilising farm water resources to grow fish or freshwater crayfish (eg yabbies). At the same time, these animals may be used to help purify water (they help dispose of effluent and they eat algae that grow on fertiliser residues).

Breeds

It is important to select the appropriate type of fish or crustacean for the climate and water conditions (ie temperature range, salinity, dissolved oxygen, purity). Permission may be

Figure 10.2 Aquaculture ponds. Growth of fish and/or crayfish is enhanced by aerating the water, in this case by circulating pond water through a spray system.

needed to grow some species in certain areas (eg trout, being an imported species, may not be permitted in some situations).

Husbandry

Most fish and crustaceans are healthier and grow faster if water contains high levels of oxygen and water quality is high.

Freshwater crayfish, trout, golden perch, Macquarie perch and Australian bass can do well in dams.

Murray cod need large, warm dams (around 1 ha or more) in warm temperate to subtropical (not tropical) areas.

A mix of golden and silver perch is more productive than either fish by themselves.

Feeding

Most fish will eat algae and insects. Supplementary feeding may improve growth rates. Crustaceans eat decaying organic matter, micro organisms and small animals.

Problems

Freshwater crayfish (eg yabbies, marron) can be cannibalistic. This is generally overcome by segregating different sized animals.

Some types of fish may not breed in a dam, but can be introduced small and grown to a large size for harvest.

In poor quality water, growth rates can be slow and susceptibility to diseases may become a problem.

Fish and invertebrates can be killed by fertilisers (particularly phosphorus and nitrogen), pesticides, animal effluent and silage.

Predators such as birds and other fish can drastically reduce stock numbers.

Cattle

Breeds

Match the cattle to the nutrient feed requirements (eg 'Black baldies' are used in Nebraska where feed is mainly grazing grass, because this breed has a relatively low protein requirement).

Some cattle breeds are more difficult to control. Brahmans, Jerseys and horned Ayrshires have bad reputations. Bulls and young steers in particular can be troublesome.

The way young animals are handled has a big impact on their behaviour in later life. Other factors that can make cattle difficult to handle include nutrient deficiencies (particularly magnesium) and windy weather.

Husbandry

- Cattle eat longer grass than sheep and are not as fussy about what they eat
- Minimise stress and you will get better meat and milk production
- Dairy cattle generally require different and better quality feeding than beef cattle
- Low input systems may not provide top quality meat production
- Male calves may be left uncastrated to reduce fat accumulation
- Cattle are sometimes kept in stalls (ie cubicles) to reduce labour and other costs, however stalls are not allowed for organic systems
- Meat quality and general health may be improved by grain feeding but this is expensive and only viable if the market is strong
- Any buildings in which cattle are kept must have optimum ventilation to minimise disease and remove ammonia from urine and bedding
- Being social animals, cattle herds need to be large enough to allow social contact, but not so large as to disrupt behaviour

Feeding

Cattle eat more at sunrise and sunset than at other times of the day.

While milking, cows need to be fed better than other cattle. This may involve supplements of hay and concentrates (eg lucerne, bran, chaff, some grain and minerals). If underfed, cows will make a lot of noise to let you know.

For top quality meat and milk production in anything but the best locations, high input supplementary feeding may be necessary.

Root crops may be an alternative to fodder in autumn and winter.

Problems

Diatomaceous earth may be given internally to reduce internal parasites.

Infertility may be affected by sodium levels in feed.

The weed 'bracken' is better controlled when sheep and cattle are grazed together, than when either are grazed alone.

Emus

See ostriches and emus

Goats

Breeds

Goats are bred and used for fleece, dairy (ie milk and cheese) and meat. They are also used for controlling weeds.

Angoras and cashmere goats are the two main breeds used for fibre production. Crossbred animals are used to provide a cross-section of production (eg meat and fibre, milk and fibre). Crossbreeding may also develop goats better adapted to regional climatic conditions.

Boer goats are a valuable breed for meat, but are also used for milk or fleece.

Dairy breeds include nubians, la mancha and saanen

Spanish goats are generally meat goats. They are very hardy, breed all year round, and require minimal management.

Husbandry

Goats prefer drier hilly country (wet conditions are not desirable). They need direct sunlight, but also shelter from excessive heat. They require protection after shearing but are otherwise hardy.

Seven goats can produce the same quantity of milk as one cow and can be maintained on a similar amount of feed and water as one cow. They are, however, only 10% the size of a cow, and easier to maintain.

Goats can be used to improve poorer farmland, grazing together with cattle and sheep. The goats eat large amounts of brush and weeds, grass then develops and cattle or sheep graze on the grass. By grazing a few goats with cattle or sheep; pasture can be protected from establishment of brush or weed species.

Feeding

Goats are more efficient ruminants than cattle or sheep, eating things which other ruminants will not. They eat mainly grasses, but will graze on a wide variety of plants. They particularly like oats, barley, lucerne (alfalfa), sunflower, linseed, corn and silage.

Problems

Moist conditions are the biggest threat. Low wet pastures can lead to lung problems (eg pneumonia). Avoid damp hay or cereal which could cause fungal problems.

Other problems may include internal parasites, Johne's disease, CAE (caprine retoviros), pulpy kidney, or attack by dogs and dingoes.

Boer goats: a sustainable breed

The recent importation of boer goats into Australia has rekindled interest in the goat meat industry. Goat meat is the most widely consumed meat in the world today even though its popularity is not in evidence in Australia.

The boer goat is predominantly a specialty meat goat that can also be used in the production of milk and angora fibre if used to cross-breed with pure angoras. One of the most positive aspects from a sustainable farming viewpoint is their strong breeding rate which ensures that stocking can replenish animals used for production.

Goats can be incorporated into a sustainable farming operation for a number of reasons including, as mentioned, meat, milk and fibre, but also as an organic weeding machine as they will devour some plant species that other stock will not touch.

It should be noted that goats can be fairly creative when it comes to escaping from enclosures and should therefore have appropriate fencing in place before their introduction. Electric fencing is viewed as an effective form of containment as long as the fence retains a live current.

If tethering a goat it is essential that a chain with swivel joints be used and this should be attached to a peg or stake that has been driven into the ground. If you use rope don't expect your goat to be there when you come back.

Horses

Breeds

The Australian stock horse is a hardy and durable horse, with a mild temperament, that is suited to harsh, dry conditions.

Clydesdales are particularly strong and sometimes used as a source of power on smaller farms.

The quarter horse has an innate ability to round up cattle, and is hardy under dry conditions.

Husbandry

Horses can be intensely managed in stables, but require feeding, watering and exercising.

Horses can be grazed in smaller areas if supplementary fed, and may be tethered as a short-term practice to control where they graze.

Fences need to be easily seen by the horse so it does not injure itself (post and rail are preferable; barbed wire should be avoided).

Avoid extreme temperatures; provide shelter in hot and cold weather.

Artificial fertilisers on pasture can inhibit pasture plants' uptake of some nutrient elements needed by a horse.

In hot or dry areas, fodder trees may be planted for shelter and supplementary feed

Feeding

Tethered horses need to be fed and watered twice each day.

Horses eat 1.5 to 3% of their body weight in dry matter each day.

If pasture quality is adequate, they will graze mainly on grasses.

A good ration needs to be balanced with adequate amounts of energy, protein, minerals and vitamins. Adequate water is also essential. If drinking hard bore water, give a ration of cider vinegar daily.

Avoid dusty feed as it can affect the horse's respiratory system. If dusty feed must be used, wet it down before feeding.

Willows may be planted as a food source in wetter, temperate areas.

Problems

Problems may include botflies, sandflies, mosquitoes, internal parasites, tetanus, colic, bighead and strangles.

Ostriches and emus

These birds are harvested for feather, meat, oil and leather. At the abattoir, every part of the bird is used. A 150 kilogram bird should yield 40–50 kg of meat.

Breeds

Proven breeds are best, though they may be expensive.

Ostriches grow up to 2.5 m tall, average 120 kg, and live 60–70 years. Hens mature after two years and reproduce for up to 40 years. Birds may be processed from 11–13 months old. Paddock maintenance is low and they can be run with other animals. Contagious disease risk is low.

Emus grow to 2 m tall and when mature weigh 35–50 kg.

Husbandry

Facilities are not expensive, but the purchase of birds can be high.

Main husbandry tasks include fencing, feeding, egg collection and incubation and rearing chicks.

Although fencing looks expensive, it need not be, due to the territorial habit of the birds. Hurricane or smooth wire fencing 1.8 m tall is normal.

Birds and food need to be kept dry for good results.

Ostriches appear to be able to be kept with cattle and sheep.

Licences are not generally needed for ostriches, but in Australia emus are protected wildlife and their farming is subject to state wildlife conservation acts.

Mature emus require at least 625 sq m per bird and juvenile emus to six months of age require at least 40 sq m per bird. Farmers frequently use the old terms for spacing, ie half an acre per breeding pair.

A breeding pair of emus can produce 20 to 40 eggs per year.

Feeding

Commercial diets are similar to other poultry.

Grit and small stones are necessary for digestion.

Pastures suitable for sheep or cattle are suitable for ostriches.

Emus are omnivorous, eating seeds, insects, pastures, fruit etc. In captivity they are often fed a mix of grain, vitamins and minerals in a pelleted form.

Problems

Ostriches are subject to various poultry diseases including mites, lice, worms, aspergillosis and bacterial infection, particularly up to the age of three months.

Leg problems in emu chicks can result in losses as high as 15%.

Food must be kept dry.

Shelter might also be needed from extreme wet weather or wind.

Predators: dogs, cats, foxes, eagles and goannas.

Lack of processing facilities is a restriction, however an ostrich/emu processing plant is established at Pyramid Hill in central New South Wales. There is also an emu processing abattoir at Alberton, near Wycheproof in Victoria.

Poultry (chickens)

Breeds

Poultry can be divided into three groups as follows:

- birds developed for egg production
- birds developed for meat production
- dual-purpose birds (both egg and meat production)

Most commercial poultry producers use hybrid (crossbred) stock for either egg or meat production, however dual purpose breeds are perhaps the most sustainable, and better for free range systems.

Dual purpose breeds tend to have the following characteristics:

- they produce good carcasses and nearly as many eggs as the light breeds
- they are placid birds that are well feathered and hardy
- when kept in deep litter houses or battery cages, they tend to become too fat and are prone to broodiness, both of which reduce their egg yield
- they nearly all lay brown or tinted eggs

Dual purpose breeds include: Rhode Island red, red Hampshire red, white wyandotte, orpington, cuckoo maran, Plymouth rock.

Husbandry

Main husbandry tasks may include fencing, feeding, egg collection, egg cleaning and fly control. Systems can include:

- Battery hens (not considered natural and generally not allowed for organic produce certification)
- Free range
- Certified organic
- Mixed with cropping (eg fruit and nuts)

In Australia, birds may be raised and marketed under a certified organic scheme if they satisfy specific requirements such as:

- Free range poultry need access to cover at all times (NASAA recommends minimum of 0.2 sq m per bird)
- Have access to outside runs at least six hours/day during daylight hours (outside areas should be at least 1 ha per 300 birds)
- Are restricted to certified parts of the farm (avoid uncertified areas such as sheds, driveways, yards, etc.)
- The wellbeing of the birds is always considered
- Artificial lighting must not increase daylight by more than 16 hours
- Litter is replaced regularly.
- Housing is disinfected with tea tree oil, steam or flame between batches

Feeding

Chickens have no appreciable sense of smell or taste, but do have a strong sense of form and colour. Substances rejected by most farm animals will be eaten by poultry. Grains with polished, shiny surfaces or brightly coloured food is eagerly consumed.

Foods that can be eaten dry but that swell when wet should be soaked before feeding and given with care to avoid the crop becoming clogged. Very dusty foods that also clog the crop and respiratory passages may be rejected by fowls.

A good poultry ration should be:

- highly concentrated with a minimum of fibre
- free from foods that might cause discomfort to the birds
- attractive in appearance and freshly mixed; damp meal becomes rancid (sour) very quickly

Birds should have access to plenty of clean water and grit. Birds kept indoors on a deep litter system will benefit from a sod of earth placed in the house. The earth is eaten readily and cleans the bowel.

Corn, sorghum and wheat grains are important in poultry rations.

Problems

Treatments for disease are usually costly and often unsuccessful, therefore preventative measures are generally preferred. Health problems include: flies, internal parasites, Newcastle disease, fowl cholera, fowl pox, infectious coryza, infectious bronchitis, Marek's disease, leucosis and coccidiosis.

Other problems may include:

- scratching the ground, destroying plants, leading to erosion if intensity is too high
- predators such as dogs, cats and foxes
- cannibalism and hysteria if birds are too highly concentrated
- unusual noises frightening birds

Pigs

Breeds

There are two main types of pigs: the pork type (for production of fresh meat) and the bacon type (for cured meat, bacon and ham).

White-skinned breeds (large white or landrace) are susceptible to sun scald and hence not well suited to free range systems.

The Berkshire, from Europe, produces a high quality carcass. Its black coat makes it more resistant to sun scald, hence more suited to extensive systems. Modern Berkshires are very distinctive with black coats and white markings. Sows are docile, good milkers and make good mothers if kept in reasonable breeding conditions.

Tamworths resist sun scalding (due to their golden red colour) and are prolific breeders. Sows are good sucklers and docile with their young. The breed is suited to open air pens or paddock feeding conditions.

The saddlebacks are black pigs with a white band over the shoulders. They are hardy, efficient grazers, able to make use of bulky fodders. They produce good sized litters of a uniform type and general high quality.

Husbandry

Three types of systems are used:

1 Pasture
2 Low-cost housing combined with pasture
3 Intensive high-cost housing (confinement) systems

Under pasture there is a need to rotate the pasture, to stop buildup of parasites.

Intensive confinement systems are used where land prices are high, but this necessitates high input and control (of pest, disease, feeding etc), and requires more sophisticated management skills. Pigs can be allowed to free range but they can cause damage. As such they are best kept in a fenced area or a pen. Pigs do not jump but they will burrow, so fencing needs to be solid.

They can tolerate severe cold or wet, but need a dry, draught-free place to sleep.

Although pigs prefer the company of others, a sow needs isolation when she furrows.

Pigs may be used to control perennial weeds in a paddock between crops, in a fenced area at a stocking rate of 25 animals per hectare.

Feeding

Pigs can adapt well to small or large properties. They are a very omnivorous animal, ie they eat a very wide variety of foods – from virtually anything humans eat, through to grass, although grass alone is not a sufficient diet. In many ways, they are perhaps the cheapest source of meat to grow.

The following are suitable crops to grow as pig feed: barley, corn, wheat, sorghum, carrots, turnips, parsnips, swedes, jerusalem artichokes, potatoes and silver beet. Pigs do need some protein, though, so foods such as these need to be supplemented with a protein supplement, such as skim or soya milk, meat meal, fish meal, cooked meat or fish, bean meal or high protein grains.

Sows in late pregnancy, and milking sows in particular, must have a good source of protein.

Pigs grown outdoors do not need mineral supplements, but if kept indoors they could require them.

Swill feeding is no longer used in Australia; it is illegal.

When fattening a pig, you can give it all it can eat until it reaches around 45 kg, after which it should be restricted to what it can eat in 15 minutes.

Pigs need pasture rich in protein (the higher the better).

Problems

Sanitation and protection from excessive heat are paramount in managing pigs; otherwise, if fed well and kept healthy, they generally present few problems.

Niacin deficiency leads to various disorders.

Heatstroke or sunburn can be a problem with pigs (avoid this by keeping shelters cool if possible – paint roofing with reflective paint and use insulation or shade trees).

Ticks (spread from dogs, bandicoots, etc) can be a problem with pigs.

Anaemia is a potential problem with piglets, and for this reason they are given an iron supplement.

Foot and mouth disease may be a problem in some countries if partially cooked meats are provided for the pigs.

Rodents, insects, mites, erysipelas, leptospirosis and lameness can also be problems.

Sheep

Breeds

Many Australian sheep are either merinos or have been bred at least partly from merinos.

The merino has been the most important breed of sheep in Australia and is bred almost entirely for wool production. Merinos are affected by their environment – the relative roles played by temperature, nutrition, humidity, light intensity, heredity, etc, are not fully understood; with respect to their influence on quality and quantity of fleece. It is known that nutrition has a very marked effect on weight and quality of fleece, together with body size and shape. There are types suited to various areas, and there is no hard and fast fixed area for the merino.

Border leicesters are less susceptible to fly strike than merinos because they are plain bodied (ie no wrinkles). They are good milkers and are mated to downs type rams to breed fat lambs or export lambs. In low rainfall areas where seasons are uncertain they are used as a dual purpose breed, the lambs finding their way to the prime lamb market in good seasons, and carried over, where possible, in lean years until they become profitable.

Southdowns produce high quality prime lambs; mutton is tender, juicy, fine grained, and of good flavour. The size is small, but compact. They are relatively slow in maturing, and are best suited to good country with short improved pastures.

The polwarth was bred specifically for the light wool growing areas found to be too wet and cold for purebred merinos. It was considered that Lincoln blood would give hardiness and weight of carcass and fleece. The polwarth may be described as resembling a plain bodied, extra long stapled merino, with less wool on the points.

Finnish landrace sheep (ie Finnsheep) are highly prolific breeders with excellent growth rates and efficient feed conversion.

Wiltshire horn sheep are a very hardy sheep breed, grown for meat rather than fleece, and particularly suited to sustainable practices. They are a short-woolled British breed that, unlike other breeds, sheds their wool each spring, leaving a hairy undercoat that provides protection from sunburn. Because this breed does not have a long fleece, the likelihood of flystrike or lice is greatly reduced, and the need for associated husbandry procedures (eg crutching, mulesing) is also reduced, if not eliminated.

The breed is relatively hardy, survives well on poor pasture, is intelligent, fertile, produces relatively vigorous lambs, and has a lean meat with a very good taste. In summary, the Wiltshire horn is an ideal meat sheep which can even convert second rate pasture into first rate lean meat, making it better suited to sustainable farming than many traditional breeds.

- Numbers of pure bred Wiltshire horns may be restricted in some parts of the world (including Australia and America) because they are not as productive as some other varieties. However, for sustainable agriculture they are valuable as they do less damage and are more hardy than many currently popular varieties.

- They are suitable for cross-breeding to produce lean prime lambs (merino crosses have little moulting).
- They are good milkers (continue to milk even on poorer quality feed), thus they may be suitable as a dairy sheep.
- They need drenching and vaccination as with other sheep, but other standard treatments (ie shearing, crutching, dagging and mulesing) are generally avoided. Rams may need treatment for flystrike on the head if they are injured while fighting.
- Having large horns, they can sometimes become caught in fences, so barbed wire is better avoided.

For more information:
Australian Wiltshire Horn Sheep Breeders Association
David Horton
Carrington Rd,
Hall, NSW 2618
Australia

Husbandry

Traditional sheep farming requires a high input of manpower and chemicals (ie pesticides and veterinary treatments). By selecting breeds such as the Wiltshire horn, these operations may be reduced, saving money and manpower, and reducing the threat of contamination.

By minimising stress you get better meat and wool production.

Feeding

Sheep graze close but can be fussy about what they eat. They cannot handle long grass, though they can be used to control certain selected weeds (eg leafy spurge).

Problems

- Worms, blowflies, ticks, lice, pulpy kidney, tetanus, blackleg, scabby mouth
- Dogs, foxes, wild pigs
- Even if the sheep is grown organically; wool may be treated with chemicals that will automatically make it unable to be certified as organic produce

Blowfly strike is a disease produced by the development of blowfly maggots on the living sheep. Susceptibility depends on climate (rainfall, temperature, humidity) and type/breed of sheep. Wrinkly sheep are more susceptible than plain-bodied sheep and merinos are more susceptible than border leicesters.

Fly strike may be controlled by breeding/selection, operations such as tail docking and mulesing (ie removing wrinkles either side of the vulva and upper surface of tail), shearing, crutching, or applying poisons.

11

Understanding products used in sustainable agriculture

It is easy to become confused by the ever-increasing range of new products being offered as an answer to sustainable farming. Many of these products are in fact very helpful, if used in the correct way and in the correct place. Because it is good, or has worked for someone else, doesn't necessarily mean that the product is suited to your farm. Some products are backed by extensive scientific research, and are proven both in the laboratory and in the field just as much as any other agricultural product. There are, however, other products which are not supported by the same degree of systematic technical development and scientific evidence. This may not mean they are not useful; it might simply mean that we do not properly (or fully) understand what they are doing (both good and bad).

Fertilisers and soil conditioners

Seaweed extracts

Seaweed contains a lot of micronutrients, plant growth hormones, sugars, alcohols, vitamins, alginates, proteins and biostimulants which improve the ability of plants to resist stresses caused by insects, the environment and disease. Pure seaweed may provide a source of nitrogen when used in bulk, but in the main it is better viewed as a plant 'tonic' or an animal feed supplement. For plants, seaweed extracts provide micronutrients, stimulate root growth, improve disease resistance, and encourage beneficial microbial development in the soil. Unless large amounts of macronutrients are added, seaweed extracts are not normally seen as a 'main fertiliser'.

For animals, seaweed extracts can be added to licks, feed supplements or used as a drench to provide mineral nutrients, improve general resistance to health problems and, in turn, reduce the need to use agricultural chemicals.

Research from the USA has shown that extracts of kelp (a type of seaweed) can:

- Increase hardiness (ie to drought and other stresses)
- Reduce some pest problems including nematodes and mites
- Increase flowering and fruit set on some fruits
- Increase yields in a range of crops including wheat, soybeans, sweet potatoes and okra
- Improve cold hardiness of some horticultural crops
- Increase fruit spurs on apples

Source: UC Cooperative Extension, 2604 Ventura Ave, Rm 100, Santa Rosa, California.

Other research has shown seaweed to reduce the occurrence of various pests and diseases, including damping off, verticillium wilt, mildew, mites, aphids, soil nematodes and cattle parasites. In addition, it has been shown to improve the effectiveness of pesticides.

Fish fertilisers

Like seaweed extracts, liquid fish extracts also contain a large variety of minerals and trace elements. Fish extracts do not contain all of the other components of seaweed (eg growth hormones), but unlike seaweed, they contain fish oil, and have a greater general effect on the inhibition of fungal and bacterial growth. Research from the USA has shown that fish emulsion can:

- promote plant growth in strawberries, soybeans, and various vegetables
- retard aging (senescence) in lettuce and peas
- delay flowering and fruiting
- reduce stress during transplanting, when used as a foliar spray
- reduce some fungal diseases by 50%

Source: UC Cooperative Extension, 2604 Ventura Ave, Rm 100, Santa Rosa, California.

Cocopeat

Peat has been used commercially in horticulture for many years. Unfortunately, it comes from a non renewable resource – peat bogs. Cocopeat, which comes from coconut fibre, is an eco-friendly alternative to mined peat moss. Cocopeat is the by-product produced when coconut husks are processed to remove the long fibres. The 'coir fibre pith' or 'coir dust' is the binding material that comes from the fibrous part of the coconut husk. It holds a lot of water and can be used as a substitute to traditional peat in soil mixes. It has a pH of 5.7 to 6.5, and a high cation exchange capacity (meaning it holds nutrients well).

Rock dusts

Rock dust is a generic term (ie it embraces a wide variety of different things that have different effects). Rock dusts are simply rocks ground into dust. Various types of rock dust are, from time to time, claimed to provide dramatic solutions to soil problems. These products vary in their characteristics, depending on where they come from. Some rock dusts do contain plant nutrients and are valuable as fertilisers (eg rock phosphate). Others contain certain micronutrients. Some rock dusts simply increase the availability of nutrients to plants. Not all rock dusts are the same!

In the home garden, mineral rock dust, including its granulated form, can be broadcast over the soil before or after planting. If changing to mineral dust as a fertiliser, remember it is slow releasing, so gradually phase out the use of other fertilisers.

If handling any of these materials, always wear a face mask to prevent inhalation of fine material which can cause symptoms like asbestosis. And always wear gloves – this is particularly important when handling perlite, vermiculite and zeolite.

The following are examples of rock dusts.

Zeolites

These are the group of minerals classified as hydrated alumino silicates. There are more than 40 different types of naturally occurring zeolites, and a lot more synthetic variants. Zeolites occur naturally in Australia, where they are mined from Mt Gipps in New South Wales. These zeolites are predominantly clinoptilolite (containing calcium, magnesium, sodium, potassium, aluminium, silicon, oxygen and hydrogen). Zeolites from other places may be different.

Plant nutrients, such as ammonium nitrogen, potassium, magnesium, calcium and other elements, have a positive electrical charge. This means that they are attracted to negatively charged surfaces in the soil profile. When there are too few negatively charged particles in the soil, nutrients are lost due to leaching. Zeolites have a honeycomb structure and strong negative charges – thus they are able to attract and hold the positively charged nutrient cations for future use by plants. This makes better use of any added fertiliser, meaning less can be applied. In cases where too much fertiliser has been applied, zeolites act as a buffer (works particularly well with sulphate of ammonia or animal manures). It is also claimed that water retention and availability are improved.

Eco-Min

Another Australian product, Eco-Min, contains important soil nutrients, including calcium, magnesium, iron, zinc, manganese, copper, cobolt, molybdenum, boron, iron and selenium. It also contains some potassium and phosphorus; but does not contain nitrogen. The producers of this product claim that it stimulates microbial activity in the soil and overcomes certain nutrient shortages. In addition, it is claimed that Eco-min improves water retention and nutrient availability in a similar way to zeolites.

Under 'normal' conditions it is recommended that Eco-Min be applied at the rate of 200 kg/ha. Due to improvements in soil conditions, when using Eco-Min it is expected that only 50–75% of normal fertiliser application is required.

Alroc Fertiliser

ALROC fertiliser is an Australian product which basically consists of 40% granite, 40% basalt, 10% dolomite and 10% bentonite. It is an organic fertiliser which can be less expensive than other conventional fertilisers. There are a number of new products now available that combine 70% ALROC fertiliser with 30% conventional fertiliser, so that conventional farmers can reduce their dependency on artificial fertilisers.

Kaolite

This is mineral containing silicon and aluminium based compounds in a 1:1 ratio. It has a very porous structure but does not disintegrate when wet. Despite containing a range of nutrients, kaolite is basically an inert material. It can be used as a soil conditioner, improving water-holding capacity and aeration; and increasing nutrient availability.

Diatomaceous earth

Diatomaceous earth is used as a natural pesticide. It is made up of the skeletons of siliceous marine and freshwater organisms. When the siliceous skeletons break up, they are like tiny particles of glass. The particles scratch through the insects' protective wax layers and they absorb some of the material. The insect loses water, dries up and dies.

For centuries before the advent of chemical pesticides, stored grain has been protected from insect attack in much of the less developed world by adding some form of powder or dust to it. Common materials include plant ash, lime, dolomite, and diatomaceous earth. There is scientific evidence that diatomaceous earth is effective as an insecticide in grain storage. Some animal producers claim it is useful in the treatment of parasites. It can be applied externally or added to feed rations (provided it is used correctly). Its use in animals is still unconventional and more research is needed to substantiate its effectiveness as a treatment for parasites.

Microbial cultures

Most plants live in association with various soil fungi which form mycorrhiza on their roots. Certain types of mycorrhizal fungi have a favourable effect on plants. For example, they may improve seed germination, root development, mineral nutrition and water uptake. Others can suppress bacterial, fungal and nematode pathogens. Commercially produced inoculants allow farmers to introduce favourable microbes into their agricultural system. Inoculants consist of the bacteria with a suitable carrier. It is critical that the inoculant contains the correct microbes for the plant being grown, and that the inoculant is fresh.

Inoculants for legumes

Legumes are widely used in sustainable systems to fix nitrogen in the soil. This ability to capture atmospheric nitrogen and 'fix' it into the soil is due to the symbiotic relationship between the legume and a bacteria called Rhizobium. In certain cases, the soil may not contain the correct bacteria and it can be useful to introduce it in the form of an inoculant.

Earthcare CLC

'Earthcare CLC' is a soil conditioner described by its manufacturers as a 'complex organic formulation based on naturally occurring humates, growth activators, amino acids, chelating agents and trace elements'. The main benefits of CLC are:

- Nutrients which exist in the soil but are unavailable to the plant (including phosphorus) are made more accessible for the plant to use
- pH fluctuations are dampened
- Lime penetrates deeper into soils
- Air, water and root penetration are improved
- Soil micro organisms are stimulated
- Rate of decomposition of organic material is accelerated
- Soil moisture is retained

Vitall fertiliser

This fertiliser was developed in the mid 1990s by chemists to minimise residues which are a problem with many other chemical fertilisers. It is produced in Australia and its manufacturers claim that it not only increases nutrient availability to plants by reducing leaching, but it also buffers the pH of the soil. It contains nitrogen (9.4%), phosphorus (2.8%), potassium (4.1%), with smaller amounts of sulphur, magnesium, calcium, boron, iron, copper, manganese and zinc.

Pesticides and insecticides

There are a large number of organic pesticides and insecticides such as pyrethrum, Neem oil, carbamates and predatory mites. These are examined in Chapter 5. It is important to remember that whilst a certain treatment may be organic, it may still have negative effects. For example, organic pesticides may not only kill the target species, but other beneficial species as well. A holistic approach is important when weighing up new products.

New plants and animals

From time to time, a new plant or animal variety emerges, offering great farming potential. The prospect of a new growth industry is often based upon some fact and some supposition. It may be that world demand for the product exceeds the supply, or that the variety offers some major advantages over current alternatives (eg a new type of sheep that is resistant to certain diseases or a type of crop that is more productive). Past examples include aloe vera, tea tree (for oil), canola (for oil) and new cotton varieties.

Usually, new varieties are a more risky proposition than a well established product – but they also offer the prospect of greater profits. When a variety hasn't been farmed to a large extent in the past, it may be difficult to obtain information on how to grow it. At the same time, markets are not established, so it can involve more work and expense to sell the product. The initial cost of purchasing stock may be quite high and there is a risk that, once stock become more widely available, an initial high investment may rapidly lose value.

A number of plants and animals are being genetically modified so that they have certain favourable characteristics, such as an ability to resist disease, or to produce a greater amount of usable materials. For example, researchers are investigating the production of microbial inoculants that are even more useful to plants – such as a microbe that has a number of beneficial effects, as opposed to just one. There is still much controversy over the development and use of genetically modified organisms in agriculture.

Biotechnology and sustainable agriculture

Biotechnology in agriculture combines biochemistry, biology, microbiology and chemical engineering to produce technologies that are useful to agriculture. Biotechnology already produces useful products such as vaccines, inoculants and microbial treatments. In fact, biotechnology is being seen by many as the saviour of modern conventional agriculture, in that it can produce plants and animals that will thrive in degraded landscapes (such as salt-tolerant species) and reduce the need for inorganic fertilisers (through the use of inoculants).

However, some experts see this thinking as similar to the revolution that occurred in the 1950s and 1960s where the development of artificial fertilisers, pesticides and herbicides was hailed as the answer to many agricultural problems. Ironically, the overuse of this technology has led to many problems in agriculture (such as acidification of soils where superphosphate has been overused). Certainly, new technologies, including biotechnology, can be useful to farmers, provided their use is sustainable both agriculturally and environmentally in the long term.

Appendix

Courses for natural gardeners

The Australian Correspondence Schools offer a range of courses by distance education that are relevant to sustainable farming. All certificate level courses and higher qualifications are internationally accredited through the International Accreditation and Recognition Council (for further information go to www.iarcedu.com). All courses are self paced, and students anywhere in the world are eligible for enrolment.

Courses include:

- The Advanced Diploma in Agriculture: core studies cover the basics of agriculture with elective studies in animal husbandry, organics, permaculture, pastures, cattle, aquaculture, and horse care.
- The Advanced Diploma in Horticulture: covers all the basics of horticulture before students choose a series of electives from turf care, arboriculture, landscaping, landscape construction, advanced propagation, irrigation, Australian native plants, herb culture, orchid culture, roses, and many others.
- The Certificate in Horticulture (organic plant growing): half the course provides a broad introduction to horticulture through subjects such as plant culture, soils and nutrition, plant identification, weeds, pests, diseases and propagation. The other half of the course deals specifically with organic growing, covering everything from soil management techniques, mulches and no dig gardening to commercial organic crop production.
- The Certificate in Horticulture (Permaculture): has a similar core content to the organic plant growing course, but devotes 50% of the course to topics specifically related to permaculture. Students who complete this course will automatically qualify for the Permaculture Design Certificate from the Permaculture Institute in Australia.
- The school conducts a wide range of short courses covering such things as resource management, permaculture, self sufficiency, organic growing, animal husbandry,

poultry, aquaculture, hydroponics, herbs, crops, ecotourism, sustainable agriculture, organic farming, environmental management, business studies and computers.

For further information contact:
The Australian Correspondence Schools
P.O. Box 2092
Nerang East, Queensland 4211
Australia
Phone: (617) 5530 4855
Email: admin@acs.edu.au
Internet: http://www.acs.edu.au

Warnborough University offers a range of degrees in natural sciences including agriculture and horticulture. These degrees are offered by distance education to students anywhere in the world. John Mason, the author of this book has been instrumental in the development of these degrees. For further information, contact:

Warnborough House
8 Vernon Place
Canterbury
Kent CT1 3WH
England
Internet: http://warnborough.edu

Further reading

John Mason, the author of this book has written several other titles that may be of interest. These are available from most good bookshops:

- *Commercial hydroponics*
- *Farm management*
- *Starting a nursery or herb farm*
- *Profitable farming*

Contacts

The following list is by no means comprehensive, but it may provide a starting point for obtaining further information or advice relating to some aspects of sustainable farming.

Alternative Farming Systems Information Centre
10301 Baltimore Ave., Room 132,
Beltsville, MD 20705-2351, USA
Internet: http://www.nal.usda.gov/afsic/

Australian Finnsheep Breeders Association
The Secretary
'The Haven', Waihemo Rd
Murringo NSW 2594
Australia
Internet: http://www.finnsheep.asn.au/

Biological Farmers of Australia
PO Box 3404, Toowoomba Village Fair
Level 1, 456 Ruthven St
Toowoomba QLD 4350
Australia
Internet: http://www.bfa.com.au/

California Certified Organic Farmers
1115 Mission Street
Santa Cruz, California 95060, USA
Internet: http://www.rain.org/~sals/ccof.html

Canadian Organic Certification Cooperative Ltd.
Certification Bureau
Box 2468
Swift Current, Saskatchewan S9H 4X7
Canada
Internet: http://www.cocert.ca

DIO (Inspection and Certification Organisation of Organic Products)
38 Aristotelous Str.
104 33 Athens
Greece
Email: cert@dionet.gr

Egyptian Center of Organic Agriculture (ECOA)
29, Yathreb Street
Dokki 12311
Cairo–Egypt
Internet: http://www.ecoa.com.eg/

Hawaii Organic Farmers Association
PO Box 6863
Hilo, Hawaii 96720, USA
Email: hofa@hawaiiorganicfarmers.org

Instituto Biodinamico (IDB)
321 – CEP
18603-970
Botucatu/Sao Paulo
Brazil
Email: ibd@ibs.com.br

International Federation of Organic Agriculture Movements
IFOAM Head Office
c/o Ökozentrum Imsbach
D-66636 Tholey-Theley
Germany
Internet: http://www.ifoam.org/

International Partners for Sustainable Agriculture
2025 I Street NW
Suite 512
Washington DC 20006, USA
Internet: http://www.igc.org/csdngo/agriculture/agr_index.htm

Kerr Centre for Sustainable Agriculture, Inc.
PO Box 588
Poteau, OK 74953, USA
Internet: http://www.kerrcenter.com/

Llama Association of Australia
Secretary
Ashby Manse Llamas
39 Blackall Rd
Batesford Vic 3221
Australia
Internet: http://llama.asn.au/

Manaaki Whenua Landcare Research New Zealand
Private Bag 92170
Auckland
New Zealand
Internet: http://www.landcareresearch.co.nz/

Nebraska Sustainable Agriculture Society
PO Box 736
Hartington, NE 68739, USA
Internet: http://www.nebsusag.org/

The Australian Ostrich Association
45 Settlement Road
Bellarine Vic 3223
Australia
Internet: http://www.aoa.asn.au/

Organic Agricultural Association of South Africa
PO Box 9834, Sloane Park, 2152
South Africa
Internet: http://www.oaasa.co.za

Organic Federation of Australia
PO Box Q455
QVB Post Office
Sydney NSW 1230
Australia
Internet: http://www.ofa.org.au/

Permaculture Institute
PO Box 3702
Pojoaque, New Mexico 87501, USA
Internet: http://www.permaculture.org/

The Emu Producers Association of Victoria
Royal Showgrounds, Epsom Rd
Ascot Vale, Vic 3002
Australia
Internet:
http://www.roseworthy.adelaide.edu.au/~pharris/EmuWeb/welcome.html#Contents

The Henry Doubleday Research Foundation
Ryton Organic Gardens
Coventry
Warwickshire
United Kingdom CV8 3LG
Internet: http://www.hdra.org.uk/

The National Association for Sustainable Agriculture Australia (NASAA)
PO Box 768
Stirling, South Australia 5152
Australia
Internet: http://www.nasaa.com.au

The Permaculture Institute of Australia
PO Box 1
Tyalgum, NSW 2484
Australia

The Rodale Institute
611 Siegfriedale Road
Kuztown, Pennsylvania 19530-9320, USA
Internet: http://www.rodaleinstitute.org/

The Soil Association
Bristol House
40–56 Victoria Street
Bristol BS1 6BY
United Kingdom
Internet: http://www.soilassociation.org

Victorian Institute for Dryland Agriculture
Private Bag 260
Natimuk Rd
Horsham, Vic 3401
Australia

Willing Workers on Organic Farms
PO Box 2675
Lewes BN7 1RB
United Kingdom
Internet: http://www.wwoof.org/

World Sustainable Agriculture Association
The President
8554 Melrose Avenue
West Hollywood
California 90069, USA
Internet: http://www.bcca.org/services/lists/noble-creation/wsaa.html

Index